T0320232

Urban Environmentalism

Whilst the global environment continues to deteriorate, cities have emerged as places of achievement and optimism. 'Cleaner and greener' cities have become a requirement of global competition and environmental protection is held to be vital to improving the lives of citizens. However, the global and urban dimensions of sustainability circle upwards and around each other like a double helix. Now into the twenty-first century, the conflictive geopolitics of the international development and collaborative urban environmental governance twist around each other with snake-like charm, and venom.

This book enquires into why cities have embraced environmental issues with enthusiasm. It locates urban environmentalism within current debates on globalization and neoliberal urbanization, and critically outlines the political success of urban environmental agendas in the postmodern condition of risk and individualization. These themes are subjected to theoretical critique and methodological exploration through Marxist analysis, discourse theory and a dialectical or relational understanding of urban environmentalism within the disruptive and often violent urban transformations of the last two decades. This approach is then applied through three in-depth second-city studies in contrasting development contexts: Birmingham (UK), Lodz (Poland) and Medellin (Colombia).

In imaginatively bringing together a wide range of disciplines, this book makes an important contribution to understanding urban environmentalism as an ideological form, operating at the levels of strategic economic interests and everyday social practices to facilitate, in place-specific ways, the legitimation of neoliberal city governments and the control/regulation of increasingly fragmented, unequal and conflictive urban societies. It will be essential reading for students of planning, geography and environmental studies, as well as to all those interested in the sociology and politics of sustainable development.

Peter Brand is Head of the Urban and Regional Planning School at the National University of Colombia in Medellin. **Michael J. Thomas** is a former principal lecturer in planning and Honorary Fellow of Oxford Brookes University.

Urban Environmentalism

Global change and the
mediation of local conflict

Peter Brand with Michael J. Thomas

Routledge
Taylor & Francis Group

LONDON AND NEW YORK

First published 2005
by Routledge
2 Park Square, Milton Park, Abingdon, Oxon OX14 4RN

Simultaneously published in the USA and Canada
by Routledge
711 Third Avenue, New York, NY 10017

Routledge is an imprint of the Taylor & Francis Group

© 2005 Peter Brand with Michael J. Thomas

Typeset in Galliard by
Florence Production Ltd, Stoodleigh, Devon

All rights reserved. No part of this book may be reprinted or reproduced
or utilised in any form or by any electronic, mechanical, or other means,
now known or hereafter invented, including photocopying and recording,
or in any information storage or retrieval system, without permission in
writing from the publishers.

British Library Cataloguing in Publication Data
A catalogue record for this book is available from the British Library

Library of Congress Cataloging in Publication Data
A catalog record for this book has been requested.

ISBN 0–415–30481–4 (Pbk)
ISBN 0–415–30480–6 (Hbk)

To Martha and Pam

Contents

Acknowledgements

Approaches to understanding the environment and its significance for the organization, construction and management of cities have evolved enormously over the last decade or so. The development of the ideas and empirical studies contained in this book grew with those changes, and we gratefully acknowledge the help of many institutions and individuals who supported the overall project and assisted in the development of specific parts.

The origins of the book lie in doctoral research undertaken at the Joint Centre for Urban Design, Oxford Brookes University. Thanks are due to many members of staff there who provided initial encouragement and critical comment, especially Roger Simmonds, Alan Reeve, Tim Marshall, Brian Goodey, Georgia Butina and Ian Bentley. The School of Planning also provided some financial assistance. Thanks are also due to the Universidad Nacional de Colombia (Medellin campus) where much of the subsequent development of those ideas took place. University research awards and assistance from the School of Urban and Regional Planning are gratefully acknowledged, as is additional support provided by the British Council and the Colombian national research body, Colciencias.

The development of the city studies required the time and cooperation of many people through help with information, exploratory conversations and formal interviews. This was not only vital to the studies themselves but also an enriching personal experience; to all, too many to mention here, our thanks and appreciation. Our thanks are also due to the anonymous referees and not so anonymous researchers who commented on earlier and partial versions published in journals and presented at conferences. Special thanks are due to the following people who read and commented on particular chapters: Mike Beazley at the University of Birmingham, Paul Watson at Solihull Council, Alan Reeve and Elizabeth Wilson at Oxford Brookes University and Tadeusz Markowski at Lodz University. Thanks also to Liliana Arboleda who prepared the maps. The research and writing of this book would not have been possible without the material and emotional support of friends and family, particularly in moments of difficulty; to Martha and Pam especially, our enduring gratitude and affection.

Chapter 3 is a revised version of an article originally published in *Environmental Planning and Management*. Part of Chapter 4 was originally presented at the *Oxford Planning Theory Conference* in 1998. Chapter 7 is based on earlier versions published in *Compact Cities: sustainable urban forms for developing countries* (M. Jenks and R. Burgess, Spon, 2000) and *Building Sustainable Urban Settlements* (S. Ramoya and C. Rakodi, ITDG Publishing, 2002).

Introduction

Making cities sustainable is now a major aim and claim of most cities in the world. Whilst international sustainable development has come up against all sorts of obstacles, city leaders have few reservations about heralding the achievements of sustainable urban development policy. The international political goodwill established more than a decade ago at the Rio de Janeiro World Summit and mobilized through all manner of sustainable development programmes, has been frustrated at every turn by the harsh realities of a deregulated global economy and, in the new millennium, pushed further into the background by shifting geopolitics and the 'war on terrorism'. Cities, it is argued, have somehow managed to escape from national and international constraints on sustainable development. Helped by de-industrialization, technological development and the rise of the service economy, cities, even in developing countries, are in many notable ways less polluted, greener and more attractive. Enormous problems still exist, of course, but cities have been able to demonstrate environmental improvements in a way denied to other spaces of the planet.

In this sense there has been a marked urbanization of sustainable development. At an obvious level, greener and cleaner cities help make them more attractive, which in turn is held to be important in the competition for inward investment and the retention of the wealthier sectors of urban populations. Cities, more than national governments, have also been keen to promote themselves as responsible actors on the world stage, conscious of and actively contributing to the solution of global ecological problems. Furthermore, in a world where high city profiles and global city networks are increasingly important in economic terms, the image of a city has become extremely important, with the exploitation of natural geographical features playing a key role in the creation of place distinctiveness. And cities and city regions, it is often argued, are less susceptible to macro-economic pressures (economy–ecology trade-offs) as well as being the spatial level at which environmental problems can be most effectively managed.

These two positions gained official currency at major international conferences at the beginning of the new millennium. Whilst the Johannesburg World Summit on Sustainable Development acknowledged the lack of progress and overwhelming obstacles to global sustainable development, at the Berlin Conference on Urban Futures cities were hailed as shining examples of the sort of progress that can be made at the environmental, economic and social levels.

In this book we examine critically some of the questions arising from these two spatially diverse and discrepant viewpoints. We are not concerned with which version is 'correct', but with how they coexist and why urban environmentalism flourishes amidst global environmental despondency. Why have cities reacted so enthusiastically to sustainability? What are the political advantages for city administrations of giving priority to the environment in urban policy when in many other fields the environment is a constraint, limitation or liability? How has this political capital been exploited in the management of cities? What sorts of processes are involved in convincing urban citizens that the environment is important to their own lives and to the well-being of the city? Is this a uniform phenomenon based on a global agenda or do cities invent their own environmental strategy? How do particular urban histories, political and development trajectories and local socio-cultural conditions affect concrete urban environmental change?

These sorts of questions signify a shift of interest away from the objective measurement of sustainability through the use of indicators, as though the environment were an external entity upon which society acts in a rational – or irrational – fashion. We argue that the significance of action on and in the name of the environment extends way beyond the impact it might have on natural resource systems. Instead we develop a relational approach based on the proposition that sustainable development proposals, and action on the environment in particular, can only be fully understood as a part of wider social change, of which the environment is neither autonomous nor necessarily the most important. From this perspective, understanding urban environmentalism needs to be seen in relation to the overall movement of society: in the transformation of the economy, forms of state intervention, social subjectivities, cultural sensibilities, and so on. This, in itself, is not a new proposition but by focusing on cities we hope to bring insights from various disciplines to illuminate the significance of space and the administration of cities, especially through the formalized processes of urban planning. We aim to put the environment in its place, as it were, both epistemologically and politically.

Critics, of course, have been keen to argue that urban environmental achievements are but a thin veneer covering the deeply unsustainable dynamics of production and consumption, and pale into insignificance in the face of the continued devastation of global ecological systems. They also point out the appalling social performance of urban sustainable development policy. Cities might look better, at least in certain sectors, but inequality and poverty have become more entrenched and profound. This serves as an important reminder of the fact that urban environmentalism emerged during a period of radical urban change involving the restructuring of economies, the decay and rebuilding of entire city sectors, major adjustments in the methods of state intervention, institutional reforms, entrepreneurial modes of urban governance, novel forms of sociality and understandings of community and new demands on citizens.

We believe that these sorts of urban transformation constitute the key to understanding urban environmentalism, and that neoliberalism is a useful and appropriate category for capturing the type of change which has been occurring over the past 20 years or so. By referring to the regulation of social change in function of the economy but exerted at the levels of ideas, institutional controls and individual subjectivities, neoliberalism provides an appropriate integrative analytic framework for understanding the dynamics of urban environmentalism. By invoking the ideological and material dimensions of change, it allows urban environmentalism to be seen as a legitimation strategy of city governments and, as such, a means to manage the inherently conflictive character of recent urban dynamics.

It follows from the above that this book is not concerned directly with the condition of natural resource systems and argues that, despite all the attention given nowadays to measurement and monitoring, any such pretence to objectivity is limited if not flawed. We hold that social discourse not scientific knowledge structures our ways of understanding and acting upon the environment. Scientific and technical knowledge form only a part of such discourse and the construction of environmental problems through political debate, policy development at multiple scales, the intersections of private interests, institutional constraints, cultural understandings, individual susceptibilities and so forth. Constructing the environment employs images and metaphors, invokes values, relates to particular temporal and spatial scales, organizes data and arguments in particular ways and calls into action particular types of intervention, thereby providing an alternative frame of reference for understanding the deficits and challenges of urban development.

Most academics (through intellectual formation) and environmentalists (through political practice) will be only too aware of the importance of the way in which discourse mediates reality, but practising planners and other professional experts tend to be much less sympathetic. This is unfortunate in at least two senses. On the one hand, planners are confronted with the intractability of environmental problems when promoting the social change on which they depend. Why solutions to environmental problems prove to be so difficult would seem to be a relevant area of reflection for planning, and the way they are constructed through discourse offers one useful and convincing explanation. On the other hand, discourse production is increasingly what planning is all about, and a critical awareness of this would also seem pertinent. The heyday of modern planning in the post-war period saw planners armed with a powerful array of public institutions and budgets of the welfare state, and the product of planning was real and tangible. Planners actually oversaw the rebuilding of the spatial reality of urban social life, and got duly criticized for their errors of judgement. In today's neoliberal state and privatized institutional environment, those instruments of planning have been blunted or discarded. Direct state intervention and generalized public funding have been replaced by partnerships, stakeholder approaches and entrepreneurial city administrations. Competition has been introduced into the heart of local government whilst privatization and deregulation have marginalized the influence and scope of planned intervention. Even the technical expertise of planners has been grossly devalued by the financial power and élan of the private sector.

As a result, planners are increasingly reliant not on the organization and application of their expert knowledge, but rather on the ability to organize urban debate and highlight issues which merit public attention and concentrate social action. Planners no longer act in the traditional sense of directing the transformation of physical environments, but rather function in a discursive environment that determines the possibilities of action of other actors. The diluted physical product of planning has been compensated by new types of discursive output, with less emphasis on statistics, land use plans and development regulations and more attention paid to more open products such as vision statements, consultative documents, competitive bidding, round table discussions, partnership projects, design guidance, place marketing, image promotion, and so on. It is only through this discursive construction of reality, present and future, that planners represent the public interest; no longer through bricks and mortar, services and infrastructure, but now words, metaphors, images, ideas and values. This communicative turn is a tacit recognition of the now predominantly discursive nature of planning.

However, the importance of discourse should not be over-exaggerated nor divert attention from the 'non-discursive' aspects of reality and the material change affecting urban societies. Discourse, as 'language in action', does not take place in a social vacuum. Much work on discourse in both the urban and environmental fields has focused tightly on discourse itself or the narrow issue which a particular discourse directly addresses, whilst paying less attention to the wider context in which meanings, problems and solutions are constructed. We see discourse not just as a means of promoting change but as a constitutive part of that change, inseparable from the dynamics of economic and social transformation implicit in urban development. This book therefore focuses on the function of discourse in the overall processes of urban transformation and examines urban environmental discourse in relation to the neoliberal reforms of the past 20 years or so. In this way, urban environmentalism is explored in its political significance in terms of the construction of new attitudes and expectations with regard to urban space, the management of urban inequality and conflict, and the legitimation of city governments; that is to say, as an ideological form.

Chapter 1 sets the scene by examining the conditions under which the environment became a central part of city policy and the administration of urban space. It outlines the weaknesses of urban sustainability rationale and policy, and argues that the explanation of such limitations lies in the process of globalization and neoliberal development. Stressing the nature of environmentalism as a social project, it is argued that the environment has been successfully constructed in such a way as to be compatible with and complement the demands made on cities by the globalization of capital, directed by entrepreneurial city governments and realized through neoliberal institutional reform. However, though impelled by the restructuring of economies, urban environmentalism has been principally directed towards the management of urban populations in a period of radical and conflictive change, ideologically displacing public concern away from the economy and on to the environment and the qualitative consumption of urban space.

Chapter 2 makes a closer and critical examination of the content and scope of urban environmental agendas. It contrasts the optimistic scenarios of urban environmentalism with the pessimism surrounding global sustainability, and explains how these two apparently contradictory visions evolved in the final decades of the twentieth century. It examines how urban environmentalism has influenced the conceptualization of space, the rationalization of planning and the introduction of a new technical agenda for cities based on natural resource protection. However, whilst natural resource protection is the express aim, urban environmentalism also incorporates enormous social claims with regard to some kind of inherent potential in the environment for improving local democracy, promoting social justice, strengthening community and encouraging better and healthier lifestyles. By demonstrating the precariousness of such claims, it is argued that the combined effect of urban environmentalism has been to reify the city so that it functions as a dispositive for modifying social behaviour in accordance with the demands of entrepreneurial urban culture and individualized urban society.

Environmental concern is not limited, of course, to cities and urban management, but it is in the lived space of cities where environmental concerns and sensibilities materialize. Arguing that scientific rationality and technical knowledge have a very limited capacity to inspire and mobilize people, appeals for changes in organizational and individual behaviour can only work to the degree that they reverberate with a particular cultural sensibility which is captured by and in the environment. Chapter 3 examines how this is achieved in the conditions of postmodernity, where the environment appears as a convincing sphere for interpreting, explaining and responding to the dilemmas of contemporary society and the challenges imposed by economic and social transformation. Risk and vulnerability, new forms of sociality, the destruction of traditional community, the reconstruction of collective identity and individual subjectivity are some of the issues which make the environment a viable axis for managing urban change. Urban environmentalism is not simply a fabrication of capital but a constitutive part of the cultural experience of contemporary social space.

The themes of the preceding chapters are brought together in Chapter 4 in a theoretical sense. It develops the conceptual resources for understanding the socially constructed nature of the environment and the solutions to environmental problems provided by orthodox environmentalism. It clarifies the role and nature of discourse in bringing about urban change, how planning theory has responded in this sense in the privatized neoliberal environment, how urban environmentalism needs to be seen as a particular form of the administration of space and its ideological nature in legitimating the transformations unleashed by the social and spatial dynamics of global capitalism.

The following three chapters develop case studies which illustrate urban environmentalism in action and show how the general propositions formulated take particular trajectories in concrete places and circumstances. Three 'second cities' from different development contexts are studied. We believe that second cities are interesting examples, more illustrative of a general phenomenon than the special case of capital cities whilst at the same time unusually creative and innovatory in terms of urban practices, in part because of their second-order status. Drawn from countries in the developed, ex-communist and developing worlds, these cities show in as clear a way as possible the different characteristics that urban environmentalism acquires; not the universal ecological rationality of world policy but the particularism and political significance of urban environmentalism. They are not comparative studies in the conventional sense of trying to demonstrate some universal law or principle through the demonstration of similarities and differences and the possible explanation of variations. Rather, the aim is to illustrate and improve understanding of how urban environmentalism operates in practice, and its significance beyond the management of natural resource systems. As a consequence, although each city study shares the same structure – organized around discourse, institutions and spatial representation – the analytic emphasis varies according to the particularities of the case in hand.

To the degree that this book aims to combine critical theoretical reflection with the analysis of practice, and general propositions with concrete manifestations, we hope that it will prove useful to researchers and students, professionals in the field of planning and other areas concerned with the management of urban change, including environmentalists working actively for better and more 'sustainable' cities. Indeed, it can be read either completely or partially according to particular interests. As for the global reach and limitations of this book, academic sources are principally European with a strong UK flavour. The case studies widen the scope of analysis through examination of cities in Western Europe, the former Communist bloc and South America, and we make ample reference to global urban environmental policy, as well as academic studies and research material from a wide variety of countries. In the sense that cities all over the world are being subject, to one degree or another, to the forces of globalization and neoliberal development strategies, we believe that the type of phenomena analysed will be relevant for those interested in urban change and environmental issues whatever the regional context.

Chapter 1

Urban change and the environment:
into the twenty-first century

Urbanization and the environment

One of the most important innovations in urban thinking in the 1990s was the introduction of the idea that cities should be sustainable. Following the Rio Summit of
1992, cities achieved a central place in the diagnosis of global environmental ills and
Local Agenda 21s were actively taken up by urban authorities throughout the world.
Sustainability became instantly a universal policy objective of urban plans, which in
turn provoked new forms of interdepartmental cooperation across city government
in an attempt to address the complex, integrative aims of environmentally friendly
urban development. These experiences were encouraged by multilateral agencies and
networked through a variety of sustainable city programmes, but relied mainly
on the initiative of local governments working closely with community and social
organizations.

This huge investment of institutional resources and community energy was
matched by considerable academic interest in the sustainable city idea. Urban theorists took up the intellectual challenge of what on the face of it appeared to be a
daunting reversal of conventional ideas and empirical evidence on the problematic
relationship between cities and ecology. Similarly, professions as diverse as architecture, civil engineering, transportation planning, landscape and urban design, public
administration and accountancy, health, and community development responded
with varying degrees of enthusiasm and conviction to the demand that their expertise should incorporate the environmental implications and effects of previously
untroubled professional practice. The environment was not only established as an
obligatory policy issue containing considerable though often conflicting public
support, it also arose as one of the most visible issues in urban politics, in the sense
of provoking high-profile protest and controversy eagerly taken up by the media.

Above all, the environment seemed to encapsulate, albeit in a confused and
contradictory fashion, the idea that urban futures could be better. After a troubled
and traumatic period of urban change in the 1980s, the environment and sustainable development appeared to offer a more harmonious path forward. After the
huge spatial upheavals and social conflict of that decade, the environment was seen
as a means to re-establish a sense of quality to urban life at community and individual levels, expressing both the practical concerns of citizens and what was left of
the intellectual optimism of academics and professionals. The environment was
holistic and inspirational, and the sustainable city arose as a new urban paradigm in
the often bleak landscape of fragmenting postmodern urbanism.

Now, into the new millennium, enthusiasm seems to have waned and hopes faded. It is already common to find academic commentators and dispirited professionals bemoaning the meagre results of years of urban environmental management. The overall trends of urban consumption and waste production have not been reversed or even significantly modified, social inequality has become more profound and the pretensions of rebuilding urban communities through an environmental conception of spatial organization has met with little success. Environmental NGOs have lost much of their ability to capture the political imagination of urban populations and influence urban policy thinking, whilst city governments plunge ever more enthusiastically into the hard struggle and potential glamour of trying to be economically competitive. The more obvious achievements of recent urban development are not to be found in the soft and subtle field of urban ecology, but in the brash and glitzy architecture of urban re-imaging, business and cultural consumption.

These real trends in urban development are reflected in the academic fields of urbanism and critical urban studies. The conceptually audacious environmental ideas and proposals of the early 1990s seem to have lost steam, now tending to drift blandly into somewhat introspective technical concerns. Certainly many of the major recent theoretical contributions to understanding contemporary urbanization can completely ignore the environment (for example, Graham and Marvin's *Splintering Urbanism*, 2001; Amin and Thrift's *Cities*, 2002; or Brenner and Theodore's *Spaces of Neoliberalism*, 2002, with a European accent) or assign it a marginal place in enquiry into new forms of spatial organization and planning (Soja's *Postmetropolis*, 2000; Castells' trilogy *The Information Age*, 1996–98; Sassen's *The Global City*, 2001, from a USA perspective). Globalization and the spatial restructuring of the world's economy has arisen as the dominant theme in urban studies and planning, with the implication that financial institutions, communications technology and informatics, along with their concomitant cultural forms, are the decisive forces behind new patterns of urbanization, relegating environmental concerns to the local, if not provincial, backwaters of urban explanation and, by implication, significance.

In this book we seek to critically review what at first sight appears to be the rise and fall of urban environmentalism, and question the assumptions behind such a general assessment. Certainly it would be untenable to write off urban environmental concern altogether. Although hard environmental achievements are scarce, it is also clear that environmental concern is now sufficiently institutionalized as to continue to exert an important influence on urban development, if only through a kind of regulatory inertia or at an aesthetic level. Equally, it might be argued that de-industrialization of the developed world's cities has in itself led to significant urban environmental improvements, although it is also widely recognized that new environmental problems continue to arise. The aggressive geopolitics of the new millennium has shifted priorities away from sustainable development and put international relations firmly back into the narrow world of trade and, more recently, the 'war on terrorism', but poverty and the environment persist as important issues on the agenda of multilateral organizations. What seems to have happened is that the anticipated articulation of economic growth, social equality and environmental protection has failed to take place, and that this is particularly evident in cities. We will question such an assumption by exploring not so much these categories in isolation, but the relations between them in the context of contemporary urban development processes.

In order to do this we will apply a range of theoretical contributions in the social and political sciences to the study of urban environmentalism. As Harvey (1996: 119) has observed, 'all proposals concerning "the environment" are necessarily and simultaneously proposals for social change and that action on them always entails the instantiation in "nature" of a certain régime of values', but for environmentalists and urban planners such change has all too frequently been regarded in a direct and deterministic manner: environmentally inspired social change should respond unequivocally to the requirements imposed by natural laws and ecological limits. However, the fact that changes of this kind have not taken place does not mean that ideas and proposals concerning the environment have failed to have any significant effect. Rather, we insist, it is necessary to examine not just the laws of ecology and the dynamics of natural resource systems but, above all, the ways in which such laws and dynamics, and the attitudes and values invested in them, have been assimilated into the overall dynamic of social and spatial transformation. In other words, we will explore the *sort* of social change that has been occurring over the past 15 years or so at an urban level and the role that the environment has played in the unfolding forms of spatial organization and urban governance. This is the general meaning we give to the term urban environmentalism.

Sustainable cities: critics and sceptics

The proposition that environmentally inspired urban change must be examined in relation to the economic, social and political contexts in which it develops is hardly new. Indeed, this is a functional requirement and viability test of sustainability's practical agenda, the importance of which has become highlighted against the paling concrete results of urban environmental management. However, at the conceptual level critical attention has tended to concentrate on the economic, social and political spheres as contextual obstacles to sustainable urban development rather than being seen as integral parts of the transformation process. The argument goes that once critical adjustments are achieved in the economic, social and political orders then improvements to the biophysical environment can both take off in themselves and consolidate social change in general, in a definitive launching of the virtuous spiral of sustainability. However, initial optimism in such a process is fading and this section briefly summarizes some of the reasons why.

First, globalization has had important consequences for the urbanization process and led to a questioning of the city as a valid unit of environmental management. The Rio Summit gave great importance to cities as centres of production (resource consumption) and pollution and therefore as key components to achieving sustainability. However, there is an increasing appreciation in environmental thinking that cities are not autonomous entities which can be made sustainable through endogenous processes of change (Low *et al.*, 2000). Globalization shifted cities from their immediate hinterlands and moved them into transnational networks and common policy frameworks, competing with each other for the footloose flows of finance capital, high-tech industries, cultural events, communication and government functions, and so forth. As a result, the global dynamics of urban development have tended to diminish and destabilize locally referenced attempts at sustainable urban development.

An early version of this argument and influential examination of the sustainable city idea was put forward by Breheny (1992). The openness of urban and regional areas was his starting point for a critique of the compact city idea then emerging in Europe. The compact city was being proposed as a high-density solution which would not only reduce natural resource consumption, pollution and waste but also recreate intensive, innovative and culturally rich urban life-settings. Breheny set out the terms of reference for much subsequent debate, in the sense of identifying a 'mixture of internal contradictions and potential conflicts with other desirable policy stances' inherent in the compact city idea. In short, he questioned the proposition that the compact city – and by inference any particular spatial form – could in itself satisfactorily resolve the multiple dimensions of sustainable development, including not only resource use and waste production but also the social and cultural values associated with sustainable urban living.

Globalization and the openness of cities have also led to a rethinking of the appropriate forms, roles and levels of government. Speculative debate on this issue explores options such as an expanded federalism (similar to the existing European Union model), extended international government (with its own juridical order), decentralist 'glocalism' and world citizenship or any number of emphases and combinations in a move towards 'cosmopolitan ecological democracy' (Gleeson and Low, 2000). It is argued that different spatial scales of government are required in order to guarantee the necessary conditions for local (urban) sustainability policy. The notion of government is also now commonly expanded by use of the term 'governance', in explicit recognition of the shift of power from the state to the private sector and hence the need to develop partnerships between government, business and social organizations in order to successfully deliver sustainable development. However, it is also recognized that partnerships involve inbuilt inequalities in the distribution of power and the consequent danger of distorting sustainable development in favour of particular interests.

Concern of this sort has led to criticism of the politics of sustainable urban development. It is argued that whilst few people would contest that sustainability is desirable, sustainability discourse tends to ignore and conceal the conflictive nature of environmental issues. Whilst sustainability addresses and tries to commit everyone, it is far from being a field of conflict-free consensus dependent only on the 'rational' acknowledgement of a common interest. The spatial conflicts which arise at the urban level are notoriously evident and difficult to overcome, for example getting people out of cars and on to public transport. If the physical environment is riddled with conflicts of interest, the social inequalities of contemporary urban development throw the political nature of sustainability into even sharper profile. In this light it is argued that protection of the physical environment should be understood as a constraint on development rather than a goal in itself. 'If the appeal for sustainability implies that only our ignorance or stupidity prevents us from seeing what we all need, and prevents us from doing it, it can undercut real reform' states Marcuse (1998: 111), arguing that social justice is the ultimate goal of progressive urban policy and politics, including those concerning the environment.

Overcoming these sorts of contextual obstacles to sustainable urban development is often held to depend on the establishment of a critical mass of sustainability conditions. The cumulative holistic change implicit in sustainability can only be achieved, it is argued, once sufficient progress has been made on a number of fronts in order to create an irreversible momentum towards new and sustainable forms of development. Many commentators acknowledge advances in technical expertise and institutional frameworks whilst remaining sceptical of their significance in the face of the underlying economic and cultural dynamics of contemporary society which severely limit the scope for change at the urban level. Blowers (2000), for example, criticizes the negative impact of centralized government and free-market policies in the UK, and the diminished position of local government and local planning to deliver urban sustainability. The key to building sustainable cities, he argues, is long-term planning, greater social equality, strong national leadership and the transformation of values in order to make sustainable change politically viable.

A final point worth mentioning at this stage is the organization of the economy itself. In Western cities criticism has centred on the power of large corporations and the business lobby to dominate urban policy and constrain and subvert urban sustainability initiatives. In developing countries criticism has been directed more at the international institutions which support global capitalism and the negative effect of global economic regulation (Burgess *et al.*, 1997) and structural adjustment programmes (Zetter, 2002) for the implementation of sustainable urban development policy. In this case, city authorities are faced with severe budgetary and institutional constraints to carry out even basic environmental health programmes in the already precarious conditions of many fast-growing cities.

Evidently such criticism of urban sustainability policy offers varied and valuable insights into the logical contradictions and practical difficulties of achieving sustainable cities. However, it does this largely within the conceptual framework of orthodox sustainability policy, which first separates then attempts to rearticulate the economic, social and environmental dimensions of development. In the following section we will examine the epistemological bases of this and alternative ways of approaching the general relationship between society and nature, and their implications for understanding urban sustainability practice.

Understanding sustainability

The purpose here is not to provide yet another exposition of the multiple interpretations of the term sustainability but, rather, to explain why its widely recognized ambiguities are an inherent quality and provide a crucial clue to its successful incorporation into current development policy and to the understanding of its applications in urban development. Criticism outlined in the previous section centred on the logical inconsistencies of sustainable development policy. In this section we aim to clarify its politics. Our concern for the moment is not so much with the explicit exercise of governmental or corporate power through legislation and international agreements but more with the struggle to establish and maintain meanings of the environment. If scientific rigour were the foundational requirement of sustainability policy and experimental demonstration a condition of its validation, it would not have survived the last century. Yet sustainability, despite the evident defects and defaults, has come to permeate all aspects of urban policy and institutional activity. Evidently, sustainability is not reliant on it being successful in its own terms, in the sense of bringing greater social equity within an enhanced and perdurable global ecology. In fact, just the opposite seems to be the case: the more the environment is revealed as a permanent cause for concern, risk and anxiety, the greater the importance of sustainable development policy.

In this sense, the political success of the environment as an issue seems to depend in part on the apparent failure to achieve any significant and lasting results. This will be discussed in more detail later, but here it should be observed that one of the important outcomes of the Rio Summit was not so much the recognition of a problem but the formalization of a solution. Scientific and pseudo-scientific versions of the depletion and possible irreversible degradation of the world's natural resource systems had been circulating for some time. Agenda 21, the policy framework for dealing with that problem, shifted the focus from diagnosis to action whilst at the same time sealing off any fundamental consideration of the nature of the problem. It developed a framework of institutional and policy initiatives for reorienting economic and social practices in favour of more environment-friendly strategies of production and consumption. The problem having been identified, what was needed was action. Acselrad (1999) describes the significance of this shift in the following terms:

> In contrast to analytic concepts oriented towards the explanation of the real world, the
> notion of sustainable development is subject to the logic of practices: it is articulated to
> the desired social effects and practical functions that the discourse pretends to turn into
> objective reality. This leads us to consider the processes of legitimation/delegitimation
> of social practices and actors. On the one hand, if sustainability is seen as something good,
> desirable and consensual, the definition which prevails will bear the authority by which
> to discriminate, in the name of sustainability, good practices from the bad. This opens
> the way for a symbolic struggle for recognition of the authority to speak about sustain
> ability. In turn, this requires the construction of an appropriate audience, an efficient field

of interlocution wherein approbation can be found. In this way one can speak in the name of (and on behalf of) those who wish for planetary survival, sustainable communities, cultural diversity, etc. The struggle for such representational authority will express the dispute over different practices and social forms which are held to be compatible with or carriers of sustainability.

(Acselrad, 1999: 36–37; our translation)

By situating sustainability in the logic of social practices rather than scientific explanation Acselrad highlights several important features. First, sustainable development is concerned with representations of the future and in this sense it is a highly political and potentially progressive proposition in that it sets out the terrain for thinking about a better future. Second, such representations are above all discursive in that they are portrayed principally through words, data, concepts, arguments and metaphors. Third, sustainability is a sphere of discursive struggle to establish a prevailing or authoritative view of what such a sustainable future might be like and how to get there. Inherent in this idea is that discourses are not abstract arguments but social practices, thus expressing interests and value systems. With this in mind we will undertake a brief review of the major approaches to understanding the environment in terms of scientific discourses or the competing epistemologies of environmentalism and their political implications.

Positivism and the environment

Sustainable development is the dominant discourse on the environment, the hegemonic representation of what the environment is and how it relates to contemporary social problems. As a development discourse focusing on action, it tends to obscure the particular understanding of nature and the environment which it encapsulates. For current purposes we will simply point out three aspects of the understanding of nature and the environment implicit in sustainable development discourse.

Separation

By dividing development into the three principal categories of economy, society and environment, sustainable development discourse continues a philosophical tradition arising in the Enlightenment, in the sense of a radical separation of mankind from the rest of the natural world. Scientific enquiry dating from Galileo and Descartes was based on the objectivization of the natural world, an analytic stance of distancing oneself from nature in order to contemplate it as an object to be studied in its own right, with its own composition, organization and dynamics. Nature was set apart and examined through new technological (telescope and microscope) and conceptual (evolution, thermodynamics) lenses, developed by a new breed of specialists (biologists, chemists, physicists). The discovery of the laws of nature brought with it a revolutionary change in the way of understanding the external world, and human life within it when scientific method was later applied through the social sciences. Much of modern Western philosophy can be seen as a response to the scientific deciphering of nature which shook the foundations of the religious cosmology of Western society constructed on the basis of a divine order integrating nature and mankind into a timeless whole.

The ecological crisis is often seen as a consequence of this philosophic stance, scientific method and resulting technological prowess. However, contemporary attitudes towards science are ambiguous. Whilst scientific knowledge is held to be a major cause of ecological degradation, the revelation of ecological problems is increasingly dependent upon it. Sustainable development discourse attempts to reinvent faith in science and technology, now as a means of reuniting the human and natural worlds. This is a difficult task as science spreads into the very codes of life through genetics, producing new ethical dilemmas and ecological dangers. However, orthodox sustainable development policy is based on a practical alliance of central government, the scientific establishment and large corporations which assume responsibility for managing such dilemmas, though in the face of mounting public concern about what this reunification of society and nature might eventually turn out like.

Instrumental rationality

The Enlightenment vision of the world opened up enormous possibilities for humanity. Knowledge, beauty and morality emerged as the autonomous spheres of science, art and religion, freeing intellectual activity from metaphysical speculation and superstition. For its part, scientific knowledge over the objectified natural world provided the practical resources to control and transform nature for the improvement of human well-being. It liberated humanity from the cyclical rhythms of nature, offered the basis for the idea of progress and hence a linear conception of time and history. Scientific rationality was 'instrumental' in the sense of providing the means to achieve human ends through, among other things, the transformation of nature. To the degree that the environmental crisis is a consequence of the transformative power of applied science, it is hardly surprising that in sustainability discourse scientific knowledge is also held to be one of the keys to its solution. Contemporary environmental problems such as climate change, ozone layer depletion, ecosystem degradation, complex forms of pollution, and so forth, defy sensorial detection and can only be described through the application of scientific expertise by specialized institutions. Just as the discovery of nature's laws was a specialist endeavour, so is the discovery of the imbalances caused by their transgression.

The idea of instrumental rationality, of using scientific knowledge to transform the natural and social worlds, carried with it the seeds of modern management. The conscious action of mankind upon its life-world introduced the need for the rationalization of the processes of change and the goals of practical action. Sustainability can be understood as a call to modify the processes of change, and hence the importance given to management in sustainable development policy. It assumes, of course, that actors should perform as rational agents, capable of adjusting their behaviour in the light of objective evidence. Sustainable development therefore places great emphasis on education, information, monitoring, indicators, state-of-the-environment reporting, exemplary practices and the exchange of experience in order to increase and circulate the appropriate knowledge base for rational, goal-oriented responses to the environmental challenge.

History and progress

The modern era arising from the Enlightenment initiated a new sense of time. The temporal notion of evolution was introduced and then became accelerated in the dizzying possibilities offered by scientific discovery. Time became a human invention as creation turned into evolution, myth into history. However, just as history was invented, so too could it be destroyed. The destructive character of much social change led to sombre reflections on the dangers and evils of a humanly produced reality throughout modern philosophy. Much of this, of course, was related to the destructive events of modern history itself.

This haunting sense of time and the perils of history permeate sustainable development discourse. The urgency of the task at hand (the 'we must take action now' imploration), along with the evident insufficiencies in terms of the results of sustainable development practice, produce echoes of disaster reverberating around the inner optimism of sustainability. Stark choices have to be made in the face of the 'irreconcilable tensions' between economic and environmental discourses (Zetter and White, 2002) or 'raw capitalism' and 'global governance' (Gleeson and Low, 2002). Doomsday scenarios of the global catastrophe always lurk in the background of even the most official reports. In a now well-established tradition, the Global Scenario Group recently described the future in terms of 'barbarization and the abyss' for a 'world in which those in power have heard the environmental warnings and are aware of the world's massive inequalities, but have ignored them in favour of profit and economic dominance' (cited in Middleton and O'Keefe, 2003: 102). Even the USA military establishment is apparently now on to it, warning that 'Climate change will destroy us Pentagon tells Bush' (*The Observer*, 2004).

Critics of sustainability within its own terms of reference therefore tend to fall into the modernist chasm opened by the separation of mankind from the rest of nature, and the logical dilemmas this produces for rational thought. For example, if economic growth, social equity and environmental protection are held to be compatible, should natural resources be given economic value, and how? Are the world's environmental resources sufficient to allow all the nations of the world to grow to the same degree? To what extent can technology overcome the finitude of natural resource availability? What is the precise relationship between poverty and environmental deterioration? How can the ethical responsibility for environmental protection be introduced into economic and social behaviour? Despite all the difficulties, sustainable development discourse puts instrumental rationality back on the counter-attack in a strenuous defence, substantiation and reassertion of the capacity of technology, and the institutions and organizations which profit from it, to uphold the idea of progress, albeit in a modified and less presumptuous manner.

Constructivism and the cultural politics of the environment

The logical positivism underpinning the sustainable development agenda of the Rio Summit has come under considerable attack, not just for its lack of success in providing real improvements to world ecology, but also for its epistemological shortcomings. Despite its claims to openness and the formal requirement that all social groups must actively participate in delivering sustainable development, sustainability is a centralized discourse heavily reliant on scientific knowledge and professional expertise not just in the natural sciences but also in law, economics, administration, and so forth. Its scientific foundations and institutionalized practices constitute a monolithic way of understanding the environment on the basis of Western philosophic tradition. This is self-evident when compared, say, to the understandings of nature in what used to be called primitive societies, but also conceals considerable variations in attitudes toward nature within contemporary society.

If the environment of sustainability discourse is a construction, what is it a construction of and what other possibilities of discursive construction exist? Language is generally considered as being important to the way in which reality is given particular meanings, and hence the enormous interest in discourse analysis as the study of 'language in action'. Discourse analysis offers a way of understanding how the use of language is associated with power and the establishment of certain meanings over other possible interpretations of events. Discourse privileges certain versions of reality over others and gives priority to certain problems and marginalizes others, establishes particular causal explanations and creates the conditions of participation in their solution, and so on. Ecological modernization as the accommodation of environmental problems to the interests of capital can very usefully be analysed in this way, highlighting how the formation of discursive alliances mobilize meanings and material resources in particular directions (Hajer, 1995). Suffice it to say at this point that discourse analysis not only opens orthodox sustainability and environmental policy to critical assessment but also politicizes issues in a new way by revealing the social biases and cultural filters within the apparently neutral and inclusionary policy discourse of sustainability.

The opening up of understandings of the environment and sustainability to the diversity of social interests and hence the multiple ways of construing a relation – both real and imagined – with nature is the aim behind cultural approaches to environmental analysis. In this perspective, the environment is not revealed through scientific enquiry nor imagined through language, but constructed through social practices which may include but are not limited to discursive ones (Macnaghten and Urry, 1998). Here, emphasis shifts to everyday experience as the key to understanding the relationship with nature. It is held that there is not a universal nature 'out there' waiting to be discovered or already imagined through language, but multiple natures construed under diverse and changing circumstances. Since everyday experience is in itself structured by complex flows of signs, goods, money, images and information, the environment becomes a 'contested sphere' not just in terms of its material exploitation but also in terms of the understandings of, responses to and engagement with what is construed to be nature and its transformation.

Such an approach attempts to extricate sustainability from the tentacles of science and enrich it with the multiple understandings arising from everyday experience. In this sense, sustainability moves beyond an interest model and participation in it would evolve beyond the mere fulfilment of preordained duties and obligations so as to 'repopulate environmental issues as they are lived, sensed and encoded in contemporary societies' (Macnaghten and Urry, op. cit.: 7). In a similar vein, Fischer and Hajer (1999: 4) argue that the sustainability metaphor is still valid but needs reinvigorating. The problem, they maintain, lies in the single dominant interpretation 'which does not compel existing institutions to reconsider the normative and cultural assumptions and premises underlying their operational practices'. This cultural approach to the environment provides a cause for renewed optimism in sustainability policy. Macnaghten and Urry (op. cit.: 3) contend that understanding the 'cultures of nature' could provide 'new and embryonic spaces for political exploration and self-discovery' through examining 'the character and complexity of human responses to nature, of people's hopes, fears, concerns and sense of engagement, and how current unease and anxiety about nature connects to new tensions associated with living in global times', and facilitate the reinvention of meaning and value lost in modernity. Similarly, Fischer and Hajer (op. cit:) aspire to politicize the cultural underpinnings of orthodox sustainable development discourse by revealing the implicit systems of meaning and frames of reference that underpin the institutional practices through which environmental politics are conducted. The institutions of sustainability would become learning institutions and reinvent themselves as 'co-producers of a new sort of development attuned to environmental constraints'.

A cultural politics of the environment offers a valuable development on the somewhat classificatory academic reflection on environmental thought, in its search for origins in the Western tradition of political philosophy or speculation on nature as a source of objective ethics for contemporary society. It reanimates the notion of cultural diversity and local identity originally proposed in sustainable development policy without falling into idealized localism or reactionary communitarianism by stressing the mediation of our relation with nature through all kinds of global flows so that social practices are now inextricably and simultaneously both global and local. What this cultural approach and the incursion into everyday life tends to underplay, however, are the structural dynamics which shape that everyday experience. It rests on an optimistic model of personal agency which constructivist analysis and much experience of sustainability policy to date throw into question.

Sustainable development and the globalization of capital

Globalization can mean many things. It might be argued that the environment was a forerunner of global thinking, to the extent that from the late 1960s the global – the planet earth – was the object of concern, whilst globalization of economic and cultural life was still in an embryonic stage. The defining characteristic of contemporary environmentalism, both in its analytic and political versions, is recognition of the global scale of the problems associated with the world's natural resource systems, involving problems which transgress national frontiers and international geopolitical divisions. The conceptual breakthrough has often been linked to the symbolic impact of the first pictures of the earth as seen from outer space, which provided a striking visual image of the smallness and fragility of the planet. Thinking globally was suddenly made easier and more important, and local environmental problems could be visualized as part of a global condition.

This early environmental thinking had only a tenuous relation to what would soon become evident as the dynamic forces of the globalization of economic and cultural life. However, the critique of capitalism as a mode of production was at the core of early environmentalism. The environmental protest movements of the 1960s were closely connected to a rejection of capitalist materialism and consumer culture, whilst radical theorists began to question whether capitalism was or could be made sustainable (O'Connor, 1994). Capitalism is, by definition, growth-oriented and expansive, and given the limited resources of the planet, the environment came to be seen as the 'second contradiction' of capitalism (the first being the exploitation of labour and class antagonism). It was argued that over the past two centuries capitalism was able to overcome periodic accumulation crises through geographical expansion (colonialism and imperialism), with the past two decades (of globalization) seeing capitalist expansion into the former Soviet bloc and the remaining remnants of the 'Third World'. Arguments in favour of a non-growth steady-state capitalist economy of the kind proposed by Daly (1996) are fundamentally at odds with the logic of capitalist accumulation, and attempts to introduce environmental externalities into accounting can only have marginal effects.

From a capitalist point of view it is argued that the only option now is environmental management. Such a proposition was at the heart of sustainable development policy and the source of so much enthusiasm at the United Nations Conference on Environment and Development (UNCED) held at Rio de Janeiro in 1992. Ten years later, at the World Summit on Sustainable Development held in Johannesburg there was not only pessimism over the chances of making capitalism environmentally sustainable but also a strident rejection of its management strategy and procedures. Overwhelming evidence was already available with regard to the continuing deterioration of natural resource systems and growing poverty and inequality, provided in considerable part by multilateral organizations and respected international NGOs. The final document was widely criticized for failing to commit to hard goals, but behind the damning empirical evidence there emerged a much stronger criticism of global capitalism and the multilateral organizations which support it.

Corporate capital had organized a sophisticated response and organizational commitment to the environment during the 1990s, for example in the form of the transnational lobbying organization Business Action for Sustainable Development, set up by the International Chamber of Commerce and its subsidiary the World Business Council, supported by most of the world's largest corporations (Rutherford, 2003). Business was now proactive in the defence of its interests and own version of sustainable development based on deregulated international trade and market solutions. However, it was the International Monetary Fund (IMF), the

World Bank and the World Trade Organization (WTO) that came under the most stinging attacks from environmental NGOs and the Left. These multilateral organizations, committed under the USA-convened Washington Consensus to the promotion of trade liberalization, privatization and the fiscal disciplining of nation states (Comeliau, 2000), were criticized for the way in which they forced governments into socially and environmentally regressive structural adjustment programmes and imposed business interests over environmental protection through trade regulation and dispute procedures (Damian and Graz, 2001). This led to energetic condemnations of 'green neoliberalism' (Goldman, 2002) and the rejection of sustainable development as a valid policy (Italian Environmental Forum, 2002) along with strong criticism of all the multilateral organizations which support it, including the United Nations and its Environment Programme (Bond, 2002).

In addition to the criticism of the growth assumption of capitalism and its environmental and social effects under current neoliberal policy, environmentalists of all political persuasions express concern over the impoverishment resulting from the marketization of nature. Under the capitalist commodity form nature ceases to have a value in its own right and is incorporated into the logic of the production of exchange value, supported by the principle of private property (see Chapter 4). This is obviously the case with regard to land, but under global capitalism it increasingly extends into the realm of water (transnational corporate take-over of water/ sewerage services and hydroelectric enterprises), air (pollution trading), biodiversity (patents on genetic resources) and ecosystems (ecological tourism and the private control of landscapes). Furthermore, the marketization of the environment leads to the material extinction of cultural diversity and the confiscation of the 'environments of survival' of the world's poor. To this degree anti-capitalist critique provides a material explanation of the sense of loss of nature that pervades much of Western culture and directly impacts on the precarious livelihoods of huge sectors of the underdeveloped world.

The marketization of sustainable development, so evident in at the Johannesburg Summit, was in fact built into its earliest footings. The Stockholm Conference on Human Development in 1972 made just one recommendation on trade and the environment in the form of the guiding principle that 'all states should . . . agree to not invoke environmental concerns as a pretext for discriminatory trade policies or for reduced access to markets' (cited in Damian and Graz, 2001: 600). The ensuing oil crises threw the industrial lobby Club of Rome's warnings on nonrenewable resource scarcity into sharp focus, and the problem was restated in terms of artificial rather than absolute scarcity due to the protectionist policies of individual nations. The ensuing action of multilateral development institutions to break down protective trade barriers and reverse models of import substitution in developing countries ensured relatively free transnational corporate access to mineral and other resources, the prices of which – with the exception of oil – plummeted on the world commodity markets to such effect that by the 1980s non-renewable resources had virtually disappeared from the sustainable development agenda. Subsequent seminal events such as the 1980 United Nations commissioned study on the relationship between development and the environment, published as *Our Common Future* in 1987, and the Rio Declaration of 1992 are underpinned by the principles of free trade, deregulation, market solutions and privatization. Once these tenets of capitalism had been established, the international financial institutions ensured they were duly respected through the regulation of both trade and national governments, thereby ensuring that environmental concerns were kept subordinate to international markets.

This is not to say that international finance institutions ignored environmental concern or turned a blind eye to the strong lobbying and political pressure exerted by both governmental and non-governmental organizations. To varying degrees the IMF, the WTO and World Bank made significant efforts to internalize environmental issues during the 1990s as part of a general move towards showing greater sensitivity with regard to non-economic issues and greater transparency in their general operation, whilst continuing to enforce the primacy of capital accumulation. Accommodation of the environment was part of a process, initiated in the 1980s, of extending loan conditioning from macro-economic policy to 'areas which were previously inconceivable such as good governance, the rule of law, judicial reform, corruption, and corporate governance' (Meyer-Bisch, 2001: 570). Part of the new openness of the IMF and the World Bank involved encouraging local civil society to participate in the assimilation or 'ownership' of the financial packages and programmes thrust upon individual countries. In the process non-governmental expertise, including that of environmental NGOs, could be publicly incorporated into their activities, with the resulting effects of the professionalization and co-optation of a large part of the environmental movement both at an international level and within the individual countries where loan conditioning operated.

These complex and often obscure processes behind the regulation of international trade and finance are insufficiently incorporated into the analysis of urban development and sustainability policy. At a general level we simply note here that sustainable development became official international development policy precisely at the moment of the globalization of capital. The World Summit of 1992 came soon after the collapse of the Soviet Union and, in the ensuing post-Cold War geopolitical climate, leaders from nearly all nations assembled enthusiastically at Rio de Janeiro. Perhaps it was not so much a shared concern for global environmental issues that provoked, as has been so often boasted, the largest ever international gathering up to that time. At least equally important was the fact that the Rio Summit provided an opportunity for the symbolic presence of all nations, large and small, in the new world order. Certainly it was more enthusiastically embraced by many countries from the developing world and the former Soviet bloc than by many developed nations. Sustainable development set the stage for a new global economic development policy, scripted in terms of a deregulated international market in which the poor countries could glimpse a significant role in the unfolding geography of global capitalist development. Their economic status remained uncertain but poor countries seemed to acquire a new importance on the basis of their 'natural capital' in an interdependent world facing unprecedented ecological challenges.

Events, of course, have proved such optimism to have been wildly misplaced. By the time of the Rio Summit virtually all nations had been put into line with regard to neoliberal economic policy. The Washington Consensus agencies were obliging developing countries to deregulate their economies, eliminate subsidies, privatize public services, reduce budget deficits, control inflation and guarantee foreign debt payments. The result was a period of economic instability typical of restructuring programmes in the less developed countries and the concentration of wealth and income distribution. For much of the developing world, the economic benefits of environmentally inspired sustainable development has added up to little more than the tiny gestures of debt-for-nature swaps and carbon trading, whilst populations collapse into poverty and political instability.

The countries involved in our case studies provide a useful illustration of the unequal coming together of nations around the environment. In Poland the environment had been the singular area of tolerated opposition under the Communist régime and the environmental movement was a key factor in its downfall, both in Poland and other Communist bloc countries. Add this political significance to Poland's own particular cultural tradition of environmentalism and it is hardly surprising that the country was one of the most energetic participants at the Rio Summit. For its part, Colombia had a history of participation in international environmental diplomacy and politics going back to the Stockholm Conference of 1972 to underwrite its active presence in Rio de Janeiro, bolstered by considerable enthusiasm in Latin America as a whole for the idea of ecodevelopment as an alternative to the import substitution development model. In contrast the UK government, out of step with much of Europe, had to scrabble together some kind of policy position at the last minute.

Globalization, neoliberalism and urban development

The globalization of capital was important not only for the construction of ideas on nature and sustainability but also for the dynamics of cities. The question arises as to how these two phenomena – the environment and urban development – have come together. It is widely accepted that the interests of capital have come to dominate sustainable development policy, creating a discourse which assimilates environmental meanings into economic development and assigning scientific knowledge an important supportive role. This is not to suggest that environmental problems have become any less or more important, but rather that they are constructed and perceived in a particular way. The scope of a revived cultural politics of the environment for challenging orthodox sustainability policy would seem to be severely limited by the global frame of reference within which people's embedded social practices are given direction and meaning. Indeed, most of the evidence to date indicates that people are reluctant or unable to adapt their lifestyles to certain basic principles of sustainable development precisely because of the rigid and demanding nature of the routines of their social experience. This suggests that the urban structuring of social experience is worth examining in some detail.

Not surprisingly, the urban implications of the globalization of capital were initially theorized by economists and urban geographers concerned with the reorganization of manufacturing production and the spatial distribution of industry. The changing industrial landscape became described as post-Fordist, in the sense of a flexibilization of mass production and consumption methods, company mergers, down-sizing and out-sourcing and the emergence of the authentically globalized processes of production, distribution and consumption, facilitated by the development of the financial and communications sectors and the emergence of the so-called informational mode of development (Castells, 1989; Amin, 1994). During the 1980s, postmodern cultural critique also began to be applied to the transforming architecture, urbanism and cultural experience of cities, remarkably synthesized and articulated to economic change by Harvey (1989). Such theorizing was aimed at capturing the enormous spatial transformations then being experienced. The massive geographical displacement of industry and the creation of a new international division of labour saw the economic base of whole cities and regions disappear virtually overnight, especially in the traditional heavy industrial areas, and the deterritorialization of local economies or their detachment from local ownership and markets. Local cultural traditions based on industry, manual work and skilled labour were truncated whilst new, non-localized cultural forms came into being in an increasingly individualized social landscape.

Globalization became the referential framework for urban development policy, and its consequences were soon integrated into the everyday experience of cities in the West and that of urban elites in countries of the former Soviet Union and the South. Globalization required cities to compete to retain existing firms and attract the footloose flows of investment capital, and the resulting urban transformations have been extensively documented: entrepreneurial government, re-imaging, gentrification, the promotion of cultural life, technology parks, convention and retail centres, business tourism and the staging of big events, and so forth. But all this has also been accompanied by widening urban inequalities in wealth and income distribution, greater spatial differentiation and class, ethnic and racial tensions.

The reorganization of urban space to overcome local Fordist accumulation crises and facilitate the insertion of cities into the new global circuits of production and consumption brought with it all the contradictions of unrestrained capitalism. It is in this context that the focus of urban studies shifted from the regulation of economic growth to the regulation of urban social life and more recent use of neoliberalism as an appropriate instrument for urban analysis. Originating in supply-side economic theory of the Chicago school in the 1970s, as an economic policy neoliberalism is most commonly associated with the Reagan–Thatcher period of the 1980s. It can be understood as the ideology of a new phase of capital accumulation (Moncayo, 2003), the linchpin of which is 'the belief that open, competitive and unregulated markets, liberated from all forms of state interference, represent the optimal mechanism for economic development' (Brenner and Theodore, 2002) and described by Bourdieu (1998) as a 'utopia of unlimited exploitation'.

For urban studies, the advantage of neoliberalism over regulation theory is its greater emphasis on government and public policy and hence the management of change at a more concrete level. Even when heralding the virtues of the market and private enterprise, neoliberalism, like any other public policy area, is enacted through the state and its politics reach out to include the systemic management of social change: the functions and operation of government, labour markets, welfare services, ideas of citizenship, and so forth. As Brenner and Theodore (2002) comment, the politics, institutional dynamics and socio-spatial effects of neoliberalism have, until recently, rarely been theorized explicitly at the urban scale. However, neoliberalism is what city politicians and planners have been wrestling with for the past 20 years or so, and it has some very specific characteristics. These can be summarized in the following terms (based on Brenner and Theodore, op. cit).

- The weakening of the nation-state as a level of economic government, the emergence of the city or city-region as a key spatial level in the global economy and competitiveness as the major principle of economic development.
- The multi-scaling of the government of cities as urban development policy is increasingly determined within a complex system of international, national, regional and local institutions.
- The restructuring of the local government, with tight fiscal controls over local authority spending and the introduction of management techniques which imitate the private sector.
- New forms of governance at the city level as local authorities realign themselves with business in the reconstruction of local economies and the promotion of an enterprise culture, partnerships in policy execution and collaborative planning.
- Radical restructuring of local labour markets, within national legislative frameworks but implemented on a local basis.
- The privatization of public services and welfare provision and the elimination of subsidies on basic service provision such as housing, health, transport and water, especially significant in developing world and transition countries.
- The rise of the service sector and the cultural economy, favouring the educated middle classes and inducing increased socio-spatial segregation, gated communities, 'archipelagos', poverty and social exclusion.

The explicit links of neoliberal urban development policy to the environment and sustainability tend to be defined in terms of:

(a) the competitiveness requirement of a clean-green city image to attract investment, leading-sector professional workers and tourists, and
(b) the demonstration of a city's sense of commitment to help solve global ecological issues through the adoption of environmental initiatives and participation in international environmental programmes.

Both are outward-looking arguments aimed at the international economy and global politics, although there is limited convincing evidence to confirm that in either case the quality of the environment is of critical importance. Indeed, industrial investment decisions may be put off by stringent environmental regulations and in any case the quality of the environment tends to be subordinate to other location decision criteria such as local labour markets, tax and other financial incentives, legislative controls over business activity, communications facilities and infrastructure provision. As a factor in determining the location preferences of high-income employees, the environment tends to lag behind other considerations such as good housing, educational opportunities for children, cultural and shopping facilities and the social status of regional cultures.

Given that cities generally have failed to make any serious contribution to global ecological sustainability and that the quality of the environment is of uncertain importance for urban economic development, it is relevant at this juncture to turn to the third dimension of sustainability discourse: the social. The critical study of neoliberalism provides a useful way of approaching this dimension in the sense that it offers an analytic perspective which goes beyond the evaluative vagueness of 'urban quality'. To the extent that neoliberal change involves the remaking of institutional landscapes for the deployment of new regulatory frameworks, it allows analysis of urban environmentalism to be inserted into the machinery and politics of the regulation of urban populations during a period of radical urban restructuring. We are thus introduced into the evolving geographies of state regulation 'in which diverse actors, organizations, and alliances promote competing hegemonic visions, restructuring strategies and development models . . . [in] an intense, politically contested interaction between *inherited* institutional forms and policy frameworks and *emergent* strategies of state spatial regulation' (Brenner and Theodore, 2002: 9, original italics). Sustainability, of course, is one of those models.

Furthermore, state regulation should not be seen as limited to the expedition of statutory or legal regulations but refers to a whole range of 'techniques of government'. On the one hand, social acquiescence requires that new regulatory measures are seen to be good or necessary, which in turn implies the creation of new social subjectivities concerning the rights and obligations of citizens and their place in the new urban order. The environment may be seen as playing an important role in this type of change. On the other hand, neoliberalism has entailed a 'dramatic intensification of coercive, disciplinary forms of state intervention in order to impose market rule upon all aspects of social life' (Brenner and Theodore, 2002: 5), including urban society's marketized relationship with nature and the environment through sustainable development discourse.

The social agenda of urban environmentalism

Since the outlining of the big issues of global sustainability at the Rio Summit, local agendas are increasingly defined in terms of organizational and individual behaviour. It is not that controversy has ceased to exist about the exact causes and possible consequences of environmental change, but rather that they have become matters for scientific expertise. Scientific controversy has become the uncertain backdrop against which non-deferrable action is played out, action being the key word from now on. The articulation of the economic, social and environmental dimensions of sustainable development, once constructed as a credible proposition, can only demonstrate the validity of its uncertain logics through the concrete actions of society upon itself. Agenda 21 was the first formulation in this respect, and since then the changing emphasis from critical reflection and analysis to action and the monitoring of its effects has been relentless.

This shift of social concern from the correct understanding of environmental problems to the mobilization of effort and action has considerable significance. The transition from analysis to activity takes the attention off causality and the structural imbalances of society and puts the spotlight not so much on effectiveness but more on the manifest intention to contribute to change on the part of the actors and agencies involved. Actors and agencies include everyone and everything, now all displayed on a homogenous plane of equal responsibility. Everyone must make a contribution according to their own activities and resources, and this contribution is measured and monitored against sustainability indicators. Performance then becomes everything. In this subtle shift of focus the heat is automatically taken off production (industry and business) which now only has to show willingness and some marginal improvement in sustainability performance. This, of course, is completely compatible with good business practice in terms of reducing production costs for example through the normal cycle of technological innovation, and allows publicity and self-promotion to replace rigorous analysis of environmental effects. In turn, the de-industrialization of cities in the Western world and the transition to service economies facilitates this process and serves to sharpen the focus on cities as centres of consumption and citizens as the primary consumer. The city is also a convenient social medium for the circulation of images and a privileged space for the demonstration of claims and superficial manifestation of sustainability behaviour.

The emphasis on the measurement of progress towards goals also has important social implications. It presupposes rational management practices and behavioural patterns oriented towards measurable achievement through monitoring and indicators. For organizations, both public and private, this managerial perspective is easily incorporated into pre-existing organizational structures. Local government sets its own targets for resource consumption and waste production in a way analogous to the technical standards imposed on private firms, and adjusts internal processes and audit procedures accordingly. However, at the urban level of collective consumption and social behaviour, the local authority has to take measures to cajole or coerce changes in general behavioural patterns. A city's environmental performance is ultimately dependent on the willingness of citizens to share and participate in the technical goals set by local government, which urban populations may neither share nor particularly care about.

It is here that urban environmental agendas take a decisive social turn. Urban sustainability policy requires the active cooperation of citizens on the practical level of everyday behaviour. This involves, among other things, the inclusion of citizens in urban management practices through such things as partnership programmes; the inducement of lifestyle changes in areas such as transport and personal health habits; government intromission into domestic life in aspects concerning energy use, eating habits and waste separation in the home; the inculcation of moral values relating to civic responsibility; the restating of the conditions of citizenship through new duties which condition access to public services; the construction of an appropriate political subjectivity no longer resting on collective organization and party representation, but the permanent fulfilment of personal obligations for 'city membership'. In this way urban environmentalism constitutes part of a new form of governmentality or the authoritative regulation of conduct towards particular objectives (Osborne and Rose, 1999), a distinctive character of which is the active and obligatory participation of citizens as political actors in carefully controlled domains and networks of urban life.

Exhortations to participate in this sort of sustainability practice are based on a rhetoric of scientific ecology and global responsibility, but the inducement of the appropriate behavioural responses is increasingly coercive in character. It is not necessarily that citizens are unaware of or indifferent to ecological issues but rather, as commented earlier, they are confronted with the contradictions and conflicts of sustainability as it affects their everyday lives. For example, how should they (we) respond to appeals to use public transport when work and social patterns are increasingly flexible and dispersed, producing demands which far exceed rigid public infrastructure provision? Why participate in local community programmes when contemporary social life is becoming ever more detached from immediate neighbourhood relations. How to reconcile personally motivated 'nimbyism' with general urban environmental interests? Environmental education and awareness programmes are inadequate in the face of such dilemmas, since they supply cognitive resources or single issue solutions which fail to account for the complexity of everyday practice. Coercion is therefore inevitable and provides the most effective thrust of the urban sustainability management strategy. Its implementation techniques are now well known: local by-laws to make non-compliance a punishable offence, the conditioning of access to services to the adoption of appropriate environmental behaviour, the imposition of sanctions such as fines and the escalation of surveillance and policing.

The urban level brings together the myriad of behavioural requirements demanded by sustainability discourse within a single spatial unit controlled by an integrative administrative structure. Whilst this, in principle, allows urban management to articulate its environmental programmes across diverse policy areas, it also multiplies the potentially manipulative quality of urban environmentalism through mutual reinforcement and the superimposition of social discipline over individual freedoms. In a pure form, this is what most environmentalists would probably argue to be needed in the neutral, holistic sense of sustainability. However, sustainability is a political programme skewed towards particular interests and invested with wide regulatory powers. In this sense environmental policy and practice can, in the name of the general good, establish a set of rules which systematically conceal private interests and condition the access and participation of different social groups in the unfolding urban order of inequality and exclusion.

Urban environmentalism and the management of conflict

The general argument developed so far has explored the proposition that sustainable development discourse and the politics of ecological modernization assure the construction of environmental problems in such a way as to minimize obstacles to continued capital accumulation. However, this widely accepted viewpoint substantially ignores the question of what positive value the environment might have to the systemic reproduction of capitalism, particularly with regard to urban change. Beyond the denunciation of 'greenwash', 'business as usual' and the 'hijacking of the environment' – which contain important political truths – to what extent can the spatialization of environmental concern be seen as contributing to capitalist urbanization in the context of globalization? We conclude this introductory chapter by summarizing the main strands of the theoretical perspective outlined so far and the analytic themes which arise from it.

First, we have described how sustainability policy operates on the cognitive strategy of separating economy, society and nature/environment into three distinct categories and then attempting to rearticulate them into a coherent and logically consistent picture of a desirable development process. Delivering sustainability then requires the building of bridges between academic disciplines, professional bodies and governmental organizations through interdisciplinary research and cross-sectoral administration. Similarly, the formulation and implementation of sustainable action demands the extension of traditional professional expertise into new fields of social policy. In turn, environmental knowledge and policy production require innovative institutional support such as interdisciplinary research groups, interdepartmental government bodies, multilateral development organizations, all interwoven into a loose network of organizations purportedly working towards the common goal of 'sustainable development'. This policy alliance has absorbed and co-opted alternative understandings of the environment and oppositional environmental politics to such an extent that refusal to join and be recognized as a legitimate partner in orthodox sustainable development now bears the cost of political and institutional marginalization. Debate and technical discrepancy are encouraged within limits, but opposition to sustainable development in any serious sense involves expulsion from a closely controlled policy circuit and retreat to the anti-capitalist, anti-globalization fringes of world politics.

At the urban level, the patchy progress of environmental management and the overall ecologically regressive trends of urbanization, together with the frustration that this has produced amongst critical expert opinion – though not officialdom – is a logical consequence of sustainability discourse itself. Sustainability is based on a flawed set of propositions whose inner inconsistency is of little concern to the power interests of those who control that discourse through the politics of ecological modernization. The demonstration of the validity of sustainable development and sustainable city policy lies in the field of social practice. When this practice fails to produce the desired results, then the apparent urgency and gravity of environmental problems are sufficient to support calls for the redoubling of efforts. The consequent emphasis on intention and action leads to a superficial preoccupation over performance, with environmental monitoring and indicators carrying the standards of expectation of organizational and individual behaviour. Criticism of sustainable urban development policy within its own logic can therefore only serve the interests of that dominant discourse by, ultimately, contributing to the call for more action of the same kind. For its part, analysis of the cultural politics of the environment introduces the diversity of social experience of the environment into what is an otherwise monolithic, flat and unnuanced version of the relationship between society and nature. In this sense it provides valuable insights into the social practices that sustainability policy seeks to modify and explains some of the difficulties involved in achieving 'sustainable' social change, without addressing the structural dynamics which drive sustainable development policy.

We maintain that sustainable development policy and the idea of sustainable cities needs to be understood within the dynamics of economic change; not economic change in general but in the specific terms of the globalization of capital and neoliberal urbanization. This is especially so in the case of cities, which have emerged with renewed force and reinvigorated functions in the networked economy of global flows. The common preamble and justification of urban sustainability policy is that half the world's population now lives in cities, but then goes on to analyse resource demands, waste production and the physical conditions of urban habitation, a functional approach which elides the particular dynamics of neoliberal urbanization. Neoliberalism as the most recent phase of capital accumulation involved huge transformations not only in the physical organization of cities but also in urban social life: local economies, work, welfare, forms of governance, identities, citizenship and sociality were all restructured. This was no smooth and peaceful transition. It involved the radical upheaval of livelihoods, political conflict, violent protest and repression, social division and changes in the law, state guarantees and individual rights. We argue that it is in this field of urban social change and conflict where the environment has played a significant role, ideologically shifting public concern from the economy to the environment and constituting a new sphere for the regulation and control of individualized urban populations.

Sustainable development has established itself as the authoritative voice in talking about the environment, despite the extraordinary amount of empirical evidence belying its weaknesses. Its strength obviously lies in its *discursive* power to shape understandings and legitimate practical action of a particular kind. It is not surprising then to find that considerable critical reflection has been directed precisely in terms of sustainability as discourse. The problem with much discourse analysis is that it limits itself to discourse as some discrete entity and ignores the 'conditions of its emergence'. Discourse is not abstract reflection but 'language in action' emerging from institutional contexts, invested with power and constitutive of social practice.

Inherent in this idea is that discursive and non-discursive practices on and in the name of the environment need to be examined in relation to the movement of society as a whole, and from our particular interest, neoliberal urbanization. This relational approach to urban environmentalism opens up alternative inroads into understanding contemporary urban government, planning and the politics of the environment. In this direction, the following three chapters will explore the urban environmental agendas arising from sustainable development policy, the environmental sensibility of contemporary society and its assimilation and activation on the part of urban planning, and the exploitation of environmental meanings in urban policy and spatial management as a governmental strategy.

In this way we will try to demonstrate that urban environmentalism, far from having had a minimal effect on urban change, can be seen as forming a constitutive part of one of the greatest transformations of space in urban history. It has been hugely successful, not in its own overt reference to ecology, but in the sense of providing argumentative and representational support for the rebuilding of cities, the spatial restructuring of urban economies, new forms and techniques of government, urban lifestyles, a sense of citizenship and political subjectivities. The theoretical resources for exploring these themes are largely Western and refer principally to societies and cities of advanced industrial countries. Globalization, it may be argued, is by definition the extension of these economic, social and cultural forms to all countries now inextricably caught up in the hegemonic world of advanced capitalism. The same observation would seem to apply to the spatiality of cities, now at least partially endowed and homogenized by a uniform display of buildings and infrastructure. However, this is only true up to a point. Globalizing forces are played out over localities with their own political trajectories and conjunctures, institutional and regulatory forms, regional cultures and spatial configurations and particular intersections of local/global relations. Underneath surface appearances lie place-specific adaptations and hybridized forms, or the path-dependency of urban change, and the implications of this for urban environmentalism will be explored principally through the three city studies.

Chapter 2
Urban environmental agendas

Introduction

After the initial euphoria of the Rio Summit and the widespread enthusiasm on the part of city authorities, spatial planners and environmental NGOs for Local Agenda 21s, formal proposals for the environmentalization of cities burgeoned. Many an urban expert saw in the environment a shift in 'tasks and mindset' comparable to the socialist-inspired impulse of post-war planning or the community-based radicalism of the 1970s (Marshall, 1994), a replacement formula for the welfare state (Palacio, 1994), the emergence of new rights and moral obligations for socio-spatial organization (Lipietz, 1996) and new possibilities to dignify city living, promote local democracy and widen the political horizons of the urban poor (Viviescas, 1993). Without paying too much attention to what sustainability might actually mean and the conditions within which it began to be pursued, environmental proposals for urban change were explored rapidly at both the conceptual and technical levels. However, it was above all a practical challenge. The policy-led character of sustainability discourse required action, and urgent action at that.

More than a decade after the Rio Summit the environment continues to be recognized as one of the major challenges of urban development and an important vector of urban planning but it no longer inspires the same optimism. Already in the mid-1990s there was growing awareness that it was far from clear what sustainability was and what it might mean in urban terms (Pugh, 1996; Jenks *et al.*, 1996), and since then critical research into urban sustainability practice has increased substantially. It has become clear that making cities sustainable not only involves modifying how they function and how they are managed, but also how problems are determined in particular ways and solutions preconfigured, and how the practical options and political horizons of urban environmental agendas are conditioned and in continual reformulation in accordance with overall development processes. Much environmental thought is based on an ahistoric conception of nature and the autonomous laws of ecology, but it became evident that sustainability policy was being moulded by particular interests and institutions. This was easy to demonstrate in the case, say, of the oil or biotechnology industries but far less obvious and apparent with regard to cities. After all, urban sustainability policy is coordinated by democratic and accountable public authorities, collectively constructed through the widespread participation of all manner of experts, social organizations and city 'stakeholders', and open to public scrutiny and debate. What was going on and going awry?

Orthodox urban sustainability policy holds that what is needed is more political will, more technological innovation, more imaginative urban management practices, and more measurement, monitoring and evaluation. This kind of argument is increasingly unconvincing and demands that urban environmental policy be examined from a more critical perspective. With this in mind, we set out to examine the environmental agendas of contemporary urban practice, simultaneously attempting to describe, analyse and criticize these agendas in their logical and practical shortcomings and explore provisionally how the limitations of the ecological modernization of cities relate to the politics of neoliberal urbanization. Before that, however, we will draw attention to the contrasting scenarios of global and urban development in the twenty-first century and trace out the policy contours of the journey which led there.

Sustainable development and cities

We have no intention of repeating the often rehearsed formal history of the sustainable development policy nor of entering into its finer details. Nevertheless, some description of the general sustainability debate is necessary to understand how it became applied to cities and in what particular ways. So let's start at the end, literally and metaphorically: at the 2002 Johannesburg Summit, the city had disappeared from the agenda. It is not mentioned at all in the Declaration, and in the Plan of Implementation it only appears through reference to already existing programmes with an urban focus. This is a curious omission given that ten years earlier at the Rio Summit cities were held to be the very key to global sustainable development.

It is easy to be cynical about the Johannesburg conference and its commitment 'to build a humane, equitable and caring global society cognizant of the need for human dignity for all'. World events both before and after the conference suggest that just the opposite is taking place. Humanism is in scarce supply, equality in a free-fall, caring hardly in evidence if a comparison of development aid and military spending is any indication (Wolfensohn, 2004), dignity has become the last outpost of an ageing generation, and the United Nations Millennium Goals for global poverty reduction, health and education, to which the Johannesburg Summit subsequently committed itself, are already looking illusory. For the purposes of our urban analysis, one particular paragraph is worth repeating here:

> We reaffirm our pledge to place particular focus on, and give priority attention to, the fight against the worldwide conditions that pose severe threats to the sustainable development of our people. Among these conditions are: chronic hunger; malnutrition; foreign occupation; armed conflicts; illicit drug problems; organized crime; corruption; natural disasters; illicit arms trafficking; trafficking in persons; terrorism; intolerance and incitement to racial, ethnic, religious and other hatreds; xenophobia; and endemic, communicable and chronic diseases, in particular HIV/AIDS, malaria and tuberculosis.
>
> (Johannesburg Declaration, paragraph 19)

This is, on all accounts, a strange mix of, not obstacles, but 'threats' to sustainable development, as if the latter already existed. Taken as a whole, the threats to the sustainable development of 'our people' seem to have acquired all the contradictory characteristics of contemporary USA foreign policy, laced with a platitudinous dose of international aid imperatives. The citation forms part of the section entitled 'Our Commitment to Sustainable Development' which underlies the substantive agenda and action areas of the Johannesburg Plan of Action, commonly referred to as WEHAB: water and sanitation, energy, health, agriculture, and biodiversity and ecosystem management. For the time being we simply observe that the Johannesburg Declaration and Plan make frequent talk of humankind, civilizations, societies, the rich and the poor, the developed and the underdeveloped, women and children, but not where they live: mostly in cities. Let us now compare this with an extract from the report following from the *Urban Futures* conference held in Berlin in 2000:

> The aim of urban policy is to produce cities which are economically prosperous, culturally vibrant, socially equitable, clean, green and safe, and in which all citizens are able to lead happy and productive lives. It follows from these objectives that cities should provide people with jobs, affordable housing and health care, education for all children, potable water, modern sanitation, convenient and affordable public transportation, nature, culture and public safety. It shall also be an objective of good urban policy to give to the citizens of every city opportunities to participate in the governance of their city and feel that they are their stakeholders.
>
> (Hall and Pfeiffer, 2000: 38)

The contrast between global vices and city virtues could hardly be more striking. Cities and their future are presented as the antithesis to the storm clouds overhanging global affairs; a utopian urban *Brave New World* which, of course, bears little relation to the real events of recent international geopolitics and the tendencies of much urban development. How did these two wildly disparate views of the world – the global and the local, the international and the urban – come to be produced and how might they help in understanding the contours of urban environmentalism?

Part of the explanation can be found in the character and timing of the conferences themselves. Both the Johannesburg and Berlin conferences announce the same general and obvious development policy aims but their references and rhetorical devices are very dissimilar. The Berlin conference was coloured by an indefatigable faith in progress and human agency, whilst at Johannesburg fear and aggression hung heavily. The intervening events of 11 September 2001 had much to do with this of course, as did the character and composition of the conferences themselves, since the former was primarily an intergovernmental meeting whilst the latter consisted of internationally recognized experts from a wide range of institutional backgrounds. Both espoused a confidence in the ability of trade and technology to overcome development problems, but whilst the former persisted in citing nation-states as the determinant geographical scale, the latter gave full recognition to the importance of cities and urban networks. The different emphasis on geographical scale in part reflects the political and professional interests of those involved. Nevertheless, the interpretation of the global–local relation is significant. Throughout the 1990s, the international optimism conveyed by the Rio Summit had collapsed into a quagmire of growing inequality, a crisis within multilateral institutions and military confrontation; cities, with their modernized and homogenized international cores, emerged as the more viable locus of international cooperation and human welfare. Rather than key centres for solving the environmental crisis, cities had become critical nodes in the organization of the global economy and operation of transnational corporations.

The second point concerns the underlying assumptions behind both the Johannesburg and Berlin conferences. Both are unrelenting in their insistence on private enterprise and market solutions. The state, it is argued, whether national or local, should be reduced to an enabling role and work in partnership with the private sector, civil society and communities, and the public ownership of companies is considered an anomaly. At the global level, the Johannesburg conference emphasizes the role of the World Trade Organization and the International Labour Organization, making careful efforts to create a sense of balance between the powers of capital and labour in the regulation of production and exchange, with national governments urged to provide appropriately deregulated frameworks and cities assigned a supportive role in making markets work more efficiently. The Berlin conference stated categorically that cities (meaning the public sector) 'are in no position to create jobs directly by running factories or service companies. Nor should they be' (Hall and Pfeiffer, 2000: 320).

At the same time, the global regulatory frameworks for neoliberal development, subsequently reiterated at Johannesburg, provide the context for urban management policy. Urban policy is determined within this neoliberal development orthodoxy and focuses on the question of 'good governance' or how cities should support markets. As Jessop (2002: 117–22) observes in his analysis of the Berlin conference, it is uncompromisingly neoliberal in its diagnosis, prescriptions and implicit aims to consolidate the neoliberal project at the urban scale through the legitimation of 'the market economy, the disciplinary state and enterprise culture'. The recipe is astonishing in both its assumptions and detail, although it is now so much part of conventional urban development policy that it is easily passed over without comment. In *Urban Future 21*, the Berlin conference effectively produced an urban policy handbook and compulsory reference work since it is policed by the International Monetary Fund, the World Bank and other multilateral development agencies. It unhesitatingly sets out what cities 'should do', the main features of which are set out in chapter 5 on 'Good governance in practice: an action plan'. Its recommendations can be summarized as follows.

• Government: cities should upgrade their administrative capacity to facilitate decentralization, free themselves from central government bureaucracy, attract strong independent entrepreneurial leaders, decentralize and democratize their internal powers, outsource public services and introduce transparent systems of contracting and performance regulation.

- Market-supportive city administration: cities should improve urban infrastructure, provide support for high-tech and no-tech, formal and informal, large and small businesses, improve land supply and property markets, streamline bureaucratic procedures, remove restrictions to growth of the business sector, introduce measures to help make markets more responsive to changing demand and changing technologies, sharpen employment-oriented deregulation, eliminate subsidies and reduce tax burdens.
- Social change: cities should promote the quality of human capital through education and training, help the economically-dependent back into productive work, promote life-long learning, protect the family threatened by rapid economic growth, promote supportive neighbourhoods in informal sectors and strengthen civil society organization, provide planning frameworks for private investment and reliable lending systems.
- Spatial organization: cities should develop flexible and demand-oriented systems of infrastructure and housing, develop ecologically-friendly transport systems, promote high-density mixed-use development, exploit architectural, urban and natural heritage to preserve unique city characters, promote polycentric city-regions composed of urban villages and garden cities, develop integrated land-use and development planning, promote urban quality and attractiveness through urban design, and assure development gain through the negotiation of packages with the private sector.
- Environment: cities should encourage reluctant national governments to develop environmental policy frameworks, develop their own eco-technologies, join eco-networks to avoid the temptation of reducing environmental standards and creating an unfair advantage in attracting investment, promote combinations of both capital-intensive and low-cost self-help environmental improvements, concentrate on the brown agenda (air, soil and water pollution) with its huge potential rewards in terms of international recognition and competitive advantage.

As Hall and Pfeiffer state (op. cit.: 321): 'Promoting economic development means more than favours to private companies. It means, among other things, providing a competitive environment, effective urban planning, an educated workforce and calculable regulations which avoid unnecessary costs'. Quite clearly, promoting economic development requires putting the whole of the city and the entire urban public policy apparatus at the service of the private sector in the process of the restructuring and regulation of neoliberal urban space. As for environmental policy, this is firmly embedded within the ethos of ecological modernization and the neoliberal re-regulation project. Absolute confidence is placed in technological innovation, and indeed a complete revolution in transport technologies by 2025, to reduce emissions and pollution at a faster rate than economic growth; both market and regulatory mechanisms are proposed, provided that regulation is uniform and guarantees a 'level playing field', and successful urban environmental management is conceived as vital to assuring city competitiveness and urban quality of life.

Returning to the disparate global–local visions of the Johannesburg and Berlin conferences, it is now clear that the Berlin proposition that cities are havens of peace, prosperity and harmony is entirely congruent with the emergence of city networks as the spatial foundation of global capitalism. With weakened nation-states and international geopolitical disorder, it is in cities that transnational capitalism can and urgently needs to legitimate itself, especially given the growing social inequalities and urban conflict. As global ecosystems continue to deteriorate and climate change emerges as a major risk, cities take the heat off the ravaging of renewable and non-renewable resources by transnational corporations and a deregulated economy, and provide a new disciplinary field of 'good governance' for public administrations and citizens. Indeed, this kind of environmentalism is now so well established, institutionalized and routinized in urban management practices that cities could be ignored in the Johannesburg-style politics of sustainable development whilst eulogized in the urban discourse of Berlin.

Getting there: a brief history of urban environmental agendas

The extension of the environmental debate to include urban issues was a relatively late development in sustainable development discourse. Cities certainly did not appear in any important way in the early seminal reports such as those of *The Ecologist* (1972), Meadows *et al.* (1972) or at the United Nations Conference on the Human Environment (1972) held in Stockholm. At the centre of early environmental concern was the destructive nature of development and the spectre of natural limits to economic growth through finite resource availability. The notion of development was opened to challenge from the capitalist centre (for example, the Club of Rome) out to the peripheries, as in the case of ecodevelopment proposals in Latin America and the idea of the need to adapt 'styles of development' to local resource bases and cultures (Sunkel and Gligo, 1980), along with critiques of capitalism's materialistic, quantitative and consumerist post-war form (Mendes, 1977; Gorz, 1978).

A decisive factor in the turn towards cities was the spatialization of demographic concern and, specifically, the urbanization of the world's burgeoning population. The population explosion became a specifically urban explosion. By the time of the Brundtland Report (World Commission on Environment and Development, WCED, 1987: 235) it could be stated unequivocally that the twentieth century was to witness a decisive urban revolution, in that 'By the turn of the century, almost half the world will live in urban areas – from small towns to huge mega-cities'. Subsequent studies were to forecast a consolidation of this urbanizing trend, so that by the year 2030 an estimated two-thirds of the world's population would be urban (United Nations General Assembly, 1995). Crucially, the 15 years between the two major early international environmental events – the Stockholm

Conference of 1972 and the Brundtland Report of 1987 – saw the re-elaboration of the environmental problematic from one of a question of *limits* to growth to that of a *condition* of development. This radical change of perspective was clearly a requirement for the full insertion of urbanization into the environmental debate since it represents, amongst other things, a shift of concern from the natural resource base to the process of development. The latter is, of course, centred on cities. It is thus only from the Brundtland Report onwards that specifically urban agendas begin to emerge from within the overall environmental debate.

The Brundtland Commission itself made five major recommendations: the development of national urban strategies to reduce urban concentration in capital cities and improve the attractiveness of intermediate-size settlements; the strengthening of local government management capacity; greater urban self-sufficiency and citizen participation; and better resource use through public–private partnerships, urban agriculture and the recycling and re-use of waste materials. Although these recommendations were global in scope, they tended to focus on the cities of the developing world. In contrast, the Report had much less to say, and was much more optimistic in tone, with regard to cities of developed countries. Whilst recognizing new environmental dangers there, it emphasized the improvements in urban environments achieved over the second half of the twentieth century, the financial and technical capacity of local administrations and the opportunities provided by de-industrialization. The report set the agenda for many bi- and multi-lateral urban programmes and recommended the development of regional programmes and city networking, research and training, cooperation with NGOs and the informal sector, and aid and technical assistance. What was still far from clear was what the city of sustainable development might actually be like. Rich in analysis, the Brundtland Report lacked an urban vision.

A tale of two cities

The Brundtland Report argued the world's population to be concentrating in towns, cities and metropolis throughout the planet. Within this overall trend, the rate of growth and sheer size of cities in the developing countries was identified as a particularly alarming problem, implying an enormous challenge in terms of infrastructure, services and housing provision. The deficiencies of Third World cities in terms of administrative power, economic resources and skilled personnel were held to be incommensurate with the scale of needs produced by rapid growth and squatter developments, worn-out infrastructure, decrepit and congested transport facilities, and poor utilities provision. Not only was the internal urban environment seen as insalubrious, with illness and disease exacerbated by high levels of air, water and noise pollution, but detrimental external effects were also held to result: uncontrolled urban growth on often scarce fertile land and the frequent contamination of surrounding natural resources and landscapes. The overall diagnosis was that: 'The world economic crisis of the 1980s has not only reduced incomes, increased unemployment and eliminated many social programmes. It has also exacerbated the already low priority given to urban problems, increasing the already chronic short-fall of resources needed to build, maintain, and manage urban areas' (WCED, 1987: 241). The urban challenge of sustainability was thus set out in terms of North and South and the different characteristics of the urbanization process in developed and developing countries. However, the objective conditions of urbanization were also to receive a quite different conceptual nuance, which may be usefully illustrated by comparing post-Brundtland thinking in Europe and Latin America.

The Latin American response to the Brundtland Report was formulated through the setting up of the Commission on Development and the Environment for Latin America and the Caribbean (UNCDELAC), established in 1989 under the aegis of the United Nations Development Programme (UNDP) and the Interamerican Development Bank (IDB). The Commission's report *Our Own Agenda* (UNCDELAC, 1990) outlined the perspective of the South on the issues raised by the Brundtland Report and, in particular, focused on the theme of the economic crisis and its relation to the environment. For Latin America, the 1980s was the 'lost decade', indelibly marked by the foreign debt crisis, deteriorating conditions of trade and an increase in poverty. In reaction to the asphyxiating debt repayment obligations and heavy economic hand of the IMF, *Our Own Agenda* sought to clarify the geopolitics of global sustainability and establish a negotiating position for the South on international development and the environment. The introduction of the notion of the 'ecological debt' of the developed world to the impoverished and ravaged South was a powerful rallying cry. However, *Our Own Agenda* added little to the discussion of urban affairs as introduced in the Brundtland Report.

Meanwhile, the optimism of the Brundlandt Report with regard to the cities of the developed world was taken up in the Commission of the European Communities' report *Green Paper on the Urban Environment* (CEC, 1990). Although new patterns of spatial organization were clearly emerging, it was confidently stated that, 'As we move towards the twenty-first century, Europe's cities will continue to be the main focus of economic activity, innovation and culture'. Three major arguments supported this position. First, the European city was considered to be endowed with sufficient economic and technological resources to be able to overcome the potential environmental damages associated with growth. Second, the European city was seen not only as an environmentally viable spatial form but also the space of individual opportunity and social cohesion. Third, the European city was seen as a central element of European civilization, its spatial and architectural heritage and cultural capital for the future.

It is here that the North–South urban conceptual divide opened by the Brundtland Report takes a clearer and more formalized shape. The European city was conceived as an object to be cherished on the basis of its 'economic efficiency, social stability, and beauty' with inherent spatial values: at once heritage, civiliza-tion and virtue in itself. In stark contrast is the Brundtland Report idea of the Third World city as a functional monster to be tamed, destitute of cultural significance, and socially and environmentally explosive. The priority in the Third World fell on the urbanization process rather than urban form, and hence the management emphasis of urban environmental programmes in the South. The outstanding feature common to both North and South was to fine-focus on issues of quality: the quality of cities and the quality of life of urban citizens.

Cities and the Rio Earth Summit

The Rio Summit, that 'momentous exercise in awareness-raising at the highest political level' (Holmberg *et al.*, 1993) only marginally advanced the discussion of urban issues. Agenda 21, the programme for implementing the principles of the Rio Declaration, attempts a comprehensive inventory of the issues relating to sustainable development, spotlighting linkages between them and formulating major lines of action. Chapter 7 of Agenda 21 refers specifically to sustainable settlements, and many other policy matters in the total of 40 chapters relate directly or indirectly to urban issues. As Haughton and Hunter (1994) observe, individual measures addressed in Agenda 21 were in themselves unexceptional; it was through their articulation in a single, internationally underwritten package that Agenda 21 provided an innovative, though complex, compendium of urban environmental policy.

The way in which cities were related to the wider development process is ideologically illuminating in itself and helps explain the general direction that urban environmental proposals were assuming. First, economic growth, as suggested in the Brundtland Report and now consecrated at Rio de Janeiro, was held to be the *sine qua non* for sustainable development. Agenda 21 reiterates free-market economic doctrine, exhorting governments to 'remove the barriers caused by bureaucratic inefficiencies, administrative strains, unnecessary controls and the neglect of market conditions', and to 'encourage the private sector and foster entrepreneurship'. This is clearly in line with the idea of urban competitiveness and the movement towards the privatization of public services being strongly promoted by the world's financial institutions. The Rio Summit also proposed a new style of 'governance for sustainable development' with three major strands: the use of market and price mechanisms to protect the environment; capacity-building in terms of institutions, personnel and technology; and a participatory approach encouraging community-driven initiatives and NGO expertise. The emphasis was on urban management, encouraging local governments to strengthen technical efficiency and work in partnership with community and business. Strict regulations cede to negotiated solutions within a flexible free-market environment. Finally, Agenda 21 urged governments to adopt national strategies for sustainability, including 'building a national consensus and formulating capacity-building strategies for implementing Agenda 21' (chapters 8 and 37) and argued that international aid, trade, cooperation and diplomacy should be increasingly conditioned by national sustainability policies in order to ensure the effective advance towards sustainability on a world scale.

Within this overall framework of neoliberal economic development policy, chapter 7 of Agenda 21 set out a specific urban environmental agenda. Promoting sustainable human settlements was defined in terms of the following eight principles:

- the provision of shelter for all,
- the improvement of urban management,
- better land use planning and management,
- the integrated provision of environmental infrastructure (water, sanitation, drainage and solid waste disposal),
- sustainable energy and transport systems,
- disaster prevention and management,
- sustainable construction technologies
- human resource development and institutional capacity building.

As mentioned earlier, there was little innovatory content in this compilation of already established UNDP and UNEP programmes, apart from an invigorated emphasis on integrated, intersectoral policies, analytic techniques such as environmental impact assessment and the use of price mechanisms.

After Rio: innovation and exemplary practice

The Earth Summit gave an important boost to environmental approaches to urban development, in that Agenda 21 formalized and structured the basic precepts and policy items of the urban environmental agenda. The period between the Brundtland Report and the Rio Summit had witnessed the production of position statements and policy documents from many multi-lateral development agencies, such as the World Bank's (1990) *Urban Policy for the 1990s*, the OECD's (1990) *Environmental Policies for Cities in the 1990s*, and the CEC's (1990) *Green Paper on the Urban Environment*. Policy statements such as these helped consolidate the urban agenda, and the challenge after the Earth Summit became one of implementation of the 'brown agenda': the pollution, risk and unhealthiness of settlements. Although recognizing the contribution of cities to the destabilization of global life-support systems (the 'green agenda' of deforestation, global warming, resource depletion, biodiversity, and so forth), the brown agenda highlighted the immediate social effects of urban environmental deterioration.

Early experience of Local Agenda 21 was to develop several key strategies of urban sustainability. First, innovation was called for in the administration of cities and management of space. Since Agenda 21 principles contain social goals, management propositions and technical challenges, the implication was that innovation had to be of an integrative type; a practical application of the holistic and ecosytemic notions underpinning sustainability, but now applied to concrete urban problems (OECD, 1996). In short, new processes were needed, using new techniques and technologies to achieve new goals, from which the sustainable city would emerge over time. Furthermore, since innovation must be adapted to local needs and opportunities, the sustainable city would emerge in a thousand different forms.

Second, the established social agenda of sustainability (health and well-being, shelter, cultural expression and political involvement) was supplemented by a stronger emphasis on social cohesion (Gilbert *et al.*, 1996; Petrella, 1996; Townroe, 1996). On-the-ground experience of the implementation of sustainable urban development came up against social fragmentation, conflict and distrust, consider-ations which the more abstract formulations of urban environmentalism had not taken into account. Social cohesion therefore arose as a vital factor in project feasi-bility. Whether in the individualized societies of the North or the crime-ridden South, it was claimed, 'the environment movement could in fact be the only hope of the human race to find a new layer of social connection, regrouping people as communities rather than self-interested and self-survival seeking alienated pockets of humanity' (Aydin, 1994: 9, speaking on behalf of the United Nations).

Third, Agenda 21 specifically assigned local authorities the role of 'educating, mobilizing and responding to the public to promote sustainable development' (chapter 28). Difficulties in the delivery of sustainable urban development gave rise to a reinvigorated concern for environmental education since implementation of Local Agenda 21 policies involved passing from a general development discourse to concrete urban practice and, therefore, the need to convince and persuade people of the practical advantages of an environmental approach. A review of local Agenda 21 experiences undertaken by the OECD (1996: 70) is adamant in this respect:

> Given the urgency of urban sustainability problems, governments cannot wait for public involvement, but must redefine public values towards ecological outcomes. Governments have found that the pursuit of sustainable development thus involves a need to actively construct 'public ideas' about sustainability. Public ideas are not necessarily convincing just because they are clearly reasoned and carefully developed. They tend to become anchored in people's minds by being connected 'with broad common-sense ideologies which define human nature and social responsibilities' (Reich, 1988, p. 79). Public ideas are important because they can generate social movements and new public norms to guide private actions.
>
> OECD (1996: 70)

Following the Rio Summit, urban environmental agendas developed a highly explicit management strategy based on community and lifestyle. In this direction Gilbert *et al.* (1996: 16) argue, for example, that 'Agenda 21 is essentially a prescrip-tion for the way people live. How people live is more a matter of what happens in communities than what happens in countries and continents'. Urban environmental agendas thus began to close down the spatial and political horizons of local social groups, especially the poor. Through environmental education governments must, the OECD exhorts, make people reconsider their individual identities and social relations in a commonsense way. Since this commonsense had to be 'constructed' from above, little room was left for autonomous and critical reflection on the causes of socio-spatial inequality and environmental problems.

The international showpiece of innovation was the Second United Nations Conference on Human Settlements held in Istanbul in 1996, a sequel to both the first conference held in Vancouver in 1976 and the Rio de Janeiro Earth Summit of 1992. Also known as the 'City Summit', Habitat II was an invitation to the world's cities to display their creative capacity and delivery-effectiveness in function of the new global policy of sustainable urban development, in which the environment had become so infused that it was a natural, almost indistinguishable part of the urban development agenda for the coming millennium (You, 1996; United Nations, 1996). Local Agenda 21 was drifting away from global ecological issues and discursively decoupling from globalization of the economy.

Johannesburg and Berlin: the double helix

The historical review of urban environmental policy undertaken so far helps to explain the disparate perspectives on sustainability as produced in the Berlin and Johannesburg conferences. From an initial common concern over the future of development, the global and local dimensions of sustainability have taken different paths that circle upwards and around each other like a double helix, each defining the trajectory of the other but connected only by the tension produced by the invisible central axis of ecology. Now, into the twenty-first century, the conflictive geopolitics of global development and the collaborative governance of urban sustainability twist around each other with snake-like charm, and venom.

The charm lies in the social agenda of urban sustainability policy. This spreads itself seductively across urban space, offering itself as 'livability' and 'quality of life', for rich and poor, in sickness and in health. The ecological management of the urban environment, integrative and protective of nature, is held to be the key to economic competitiveness, good government, social cohesion, healthy communities and individual well-being. Differences within cities and between cities of the North and South are erased in this globally concerted effort to produce sustainable world development through new urban practices.

The urban environmental agenda also absorbs new global sustainability problems and priorities with ease, whether social, ecological or political. Intensified global poverty, for example, has adjusted early environmental concern over the urbanization of the world's population to a new formulation concerning the urbanization of the world's poor. In this direction, the UN-Habitat programme takes pains to re-urbanize global sustainability policy after Johannesburg by articulating water, sanitation and settlements as an 'integral cluster', focusing on the problem of slums (UN-Habitat, Kajumulo, 2004). Similarly, the emergence of new global ecological threats can rearrange priorities and arguments with, for example, growing concern over climate change and water compensating flagging interest in other areas (White, 2002). Furthermore, good environmental governance is held to rest on universal principles as old tensions between East and West, North and South, dissolve in the globalized world of networked cities. Such governance and its accompanying spatial innovation are, of course, based on market principles and compliance with the requirements of international capital.

In the following sections we will develop a critical account of urban environmental agendas arising largely from the Rio Summit and given full expression at the Berlin conference. Even leaving aside national government policy variations, at first sight this is a daunting task since there is now a profusion of proposals from multilateral agencies, academics, professional bodies and environmental NGOs. Although always addressing a range of issues and often consisting of compendia, emphases vary enormously, from planning and management (for example Blowers, 1993; Haughton and Hunter, 1994; Miller and de Roo, 1999; Lein, 2003), nature and ecology (Girardet, 1992; Hough, 1995; Register, 2002; White, 2002), urban form and design (McHarg, 1969; Moughtin, 1996; Jenks *et al.*, 1996; Thomas, 2002), social and community issues (Roseland, 1998; Barton and Tsourou, 2000; Carley *et al.*, 2001), sanitation and health (Barton, 2000; McGranahan *et al.*, 2001) and technological concerns such as energy and transport (Samuels and Prasad,1994; Newman, 1999; Banister, 2000; Hoogma *et al.*, 2002). Furthermore, different countries or regions often develop particular approaches, such as bioregionalism and community in North America (Aberley, 1994; Roelofs, 1996; Register, 2002), as well as hybridized approaches for developing countries (Satterthwaite, 1999; Pugh, 2000; Jenks and Burgess, 2000; Hardoy *et al.*, 2001).

The task is made easier by the fact that urban environmental agendas have acquired a distinct universality. A basic consistency is established through formalization in international declarations (such as Aarlborg and Istanbul), coordination by multilateral organizations (governmental and non-governmental), and implementation via international programmes and city networks such as the Sustainable Cities Programme (UNEP/UN-Habitat), the Healthy Cities Programme (WHO), the Urban Management Programme (UNDP/World Bank/Habitat), the CITIES Programme (European Commission), as well as more independent associations of cities such as the International Council for Local Environmental Initiatives (ICLEI) and Citynet (Asia). Additionally, agendas tend to be based on, or at least referenced to, scientific knowledge (or knowledge disputes) and promoted by experts who move freely between academic research, government, consultancy and environmental activism of one sort or another, thereby establishing intense relations and frequent exchange across institutional borders. In other words, there exists a closely-knit policy community which establishes strategic boundaries, cognitive maps and practical priorities. We do not, therefore, propose a systematic analysis of a representative selection of agendas but, rather, to undertake a thematic analysis of the topics and fields of contention through which such proposals are formalized, substantiated, subjected to analysis, projected and experimented in practice.

In so doing we will also try to identify the 'venom' of urban environmentalism through political critique. As we have persistently pointed out, urban environmental agendas emerged over a period of global restructuring of capitalism following the crisis of Fordist production and the Schumpeterian welfare state (Amin, 1994; Jessop, 2001) and unfolded in the re-regulatory phase of neoliberalism. During the early stages, the environment was both a material obstacle and focus of cultural opposition to capital accumulation in a general sense, and by implication at least was anti-urban. Now, into the twenty-first century, urban environmentalism is celebrated by neoliberal governments at all scales. In this context, we will try to explore to what extent the environmental governance of urban space has become implicated in globalized capital accumulation and the management of its social contradictions and specifically urban conflicts.

Cities, space and planning

The attempt to harmonize urban development with nature and integrate the dynamics of urbanization with natural resource systems led to new conceptualizations of the city and novel directions in urban planning. The sustainable city idea brought to the surface and gave life to what had previously been the closeted concerns of engineering, thereby requiring a radical revision of the analytics of space and the rationality of planning. The functionalist assumptions of policy thinking deriving from the Rio Summit make it of limited value for exploring this aspect and it has been the task of urban experts to give conceptual form to the sustainable city idea.

The conceptualization of cities

The Brundtland Report established a clear if not causal relation between cities and the global ecological crisis, in terms of the urbanization of the world's population and the concentration of resource consumption and waste production in cities. In subsequent urban environmental thinking this was given the character of a *fait accompli* of evolutionary naturalness. Thus, for example, Girardet's (1992: 11) opening sentence runs: 'Sometime around the turn of the millennium an urban baby will be born whose birth will tip the balance statistically, for humanity, from being a predominantly rural into a predominantly urban species', and Lester R. Brown comments in the book's foreword that, after millions of years as a hunter and gatherer, and thousands of years as a cultivator, mankind has urbanized in the space of only a few centuries, especially the present one, and comments that '[i]n evolutionary terms, this shift has occurred at a breath-taking pace, creating problems never before encountered'. Equally, the evolution of cities is held to represent a crisis of civilization, as when Rogers and Gumuchdjian (1997: 4) argue the paradox that cities, the birthplace of civilization, are now the major destroyer of ecosystems and the greatest threat to human survival on our planet. Statistical and conceptual appreciations of this kind now preface most works on sustainable cities, frequently emphasizing regional growth differentials and the disproportionate growth of megacities in the developing world. Broad-brush statements of this kind sweep away the need for detailed geographical–historical analysis and attention to the processes of urbanization.

Having established this general relation, other arguments are commonly used to give it substance. First, cities are seen to be an ideal unit for local action to resolve global problems since, it is argued, they constitute a network of centres of information and institutional power in the new world economic order. Second, cities are held to be an effective unit of environmental policy implementation, management and innovation since they possess administratively definable spatial limits and a manageable scale which facilitate the practical comprehension of difficult environmental problems. In this sense city-scale government allows, in principle at least, for integrated policy responses to the trans-sectoral nature of environmental problems (for example, industry, transportation and air pollution) and a viable management approach through land use planning. Thus the UN-Habitat centre (2004) holds that 'in human settlements . . . actions have to be coordinated and managed. It is at this level that policy initiatives become an operational reality and an eminently political affair. It is here that local actions must and can deliver global goals'. Third, cities are conceived as reliable depositories for optimism. In part this derives from the previous two points: if cities are the focus of environmental problems and at the same time viable and effective centres for environmental management, then there is an in-built bias for believing that cities can contribute to sustainable development. Much initial thinking was pitted against social and economic decline and looming environmental catastrophe, with early urban environmental agendas adjudicating a 'missionary role' to cities (Nijkamp and Perrels, 1994). Rogers and Gumuchdjian (1997) reflect a more recent style of optimism, citing ecological enlightenment, communications technology and automated production as the fount of urban sustainability in the twenty-first century.

These sorts of argument produced novel ways of understanding the city as a spatial object and for prescribing urban futures. The most radical stance is to define the city in terms of concepts borrowed directly from the natural sciences, especially ecology. Girardet (1992: 16) exemplifies this approach when he describes cities as 'organisms with an evolutionary history, a life of their own', which can be understood in terms of 'their metabolism, their communities, their climate, and their environs'. Within such a conceptual framework, cities are inevitably devourers of natural resources with a 'parasitic' relation to their 'host environment', and the urban challenge consists of establishing a symbiosis with the natural world. In this line of argument cities must become 'benign organisms' through the establishment of circular metabolisms that imitate the functioning of natural ecosystems.

Alternatively, the city may be seen as analogous to a natural entity for analytic purposes. Thus, for Nijkamp and Perrels (1994: 10–11), 'Urban (or metropolitan) regions are essentially large production and information processing systems . . . Like all competitive dynamic systems, urban systems may be regarded as biological species, evolving with different growth rates and in different directions'. Within this evolutionary frame of reference, cities are seen to pass through stages of development, similar to ecosystem dynamics, and in constant competition for survival with other members of the urban 'species'. Similarly, White (2002: 5) argues that to see cities as if they were ecosystems operating within the biosphere not only privileges the interactions between urban and ecological systems but also 'makes it easier to think in terms of everything being linked to everything else, as in natural systems'.

With the ahistoricity of systems theory as the main conceptual device, along with its apparent neutrality as an analytic tool, urban environmental agendas can be constructed without undue concern over structure and process. This new function-alism privileges the problem of behaviour (of individuals and organizations) and gives priority to solutions in terms of management, technology and ethics. Urbanization is taken as an evolutionary phenomenon and the significance of the globalization of capital is subsumed within Darwinian analogies of urban competi-tion. This does not mean that the market is ignored, since this is assumed to be as natural as the production of urban space. Rogers and Gumuchdjian (1997), for example, concedes that the construction of cities is in the hands of market forces and dictated by short-term financial imperatives with often chaotic results, but ignores the fact that this might have something to do with the logic of capital, holding that it is simply a challenge to political will and vision and an opportunity to create a new enlightened citizenry. Sustainable urban development is held to be a question of 'choices' which 'can and need to be made, not just by governments but also by individuals in the conduct of their lives' (White, 2002). For the rational-choice individual, instruments are now available to assist in appropriate decision-making through a personal ecological footprint calculator. For the city, sustainable develop-ment is now associated with 'smart growth' and being not only clean and green but also technologically advanced, enterprising and attuned to the market.

The rationalization of planning

It has been widely argued that sustainable development offered renewed horizons for local planning. As an activity concerned with the rational use of land in the name of the public good, city planning came under aggressive attack from the more radical versions of the neoliberal deregulatory project. Although the situation varied consid-erably between countries, the general trend was toward market-based decisions on investment, the spatial distribution of activities, and building and infrastructure provision, all of which seriously undermined the planning profession's claims and powers to influence urban development. This was barely compensated by emerging forms of more horizontal urban governance since the modernist planner was marginalized by city leader partnerships with local business elites whilst at the same time having increasingly little to offer the ordinary citizen in terms of spatial welfare. Stranded between entrepreneurial governance and a contracting state apparatus, city planning offices were in danger of becoming a mere appendage to the urban devel-opment process. Then along came the Rio Summit arguing that local authorities were important and *the* key players in delivering sustainable development, calling on them to consult local society in the preparation of consensual plans for sustain-able development. With Local Agenda 21, planning found a new practical function (Mittler, 2001) in which every settlement, the ordinary and provincial as well as the glamorous and global, had an equally important role to play.

Our interest here is not to assess practical outcomes but to outline the impli-cations of sustainability with regard to the rational foundations of planning as a state activity and its integration into planning's theory of itself. There is little doubt that spatial planning has undergone radical transformations in the postmodern period.

In a European context at least, contemporary planning has been described as a more open and complex system characterized by its enabling role within new forms of governance, its communicative rather than positivist rationality, its adoption of the entrepreneurial ethic and the shift from equality to inclusion as the primary social goal (Tewdwr-Jones, 2001). This general description of how planning sees itself is increasingly valid for cities throughout the world of global capitalism and neoliberal state policies, and expresses the state's subordination to the private sector in economic development, the assimilation of business management procedures by local authorities and a radical change in the social function and institutional practices of planning itself.

Theoretical accommodation of such change was to a large extent provided by a communicative theory of planning (Forester, 1993; Healey, 1996). The long-standing professional goal of social equality was no longer tenable within the neoliberal state apparatus compromised by business interests, the redesign of welfare and the socially divisive effects of urban economic restructuring. Forced into the abandonment of a central command role, professional planning began to see itself as fulfilling an essentially communicative function: establishing dialogue, achieving consensus, clarifying roles and responsibilities, and collaborating in the participatory definition of goals, performance targets, management practices, monitoring procedures and budget allocations. As the planning profession began to redefine itself as managing communicative practices rather than state budgets and public investment, now part of a new régime of institutions and actors collectively organizing social change, theoretical reflection tended to coalesce around discourse theory and its many variants for understanding the nature and role of planning practice.

City planning, like any other state activity, was realigned to the demands of capital accumulation and neoliberal regulatory practices in the restructuring of urban economies, and whilst this did result in some impressive urban and architectural transformations it also had enormous social costs. As whole sectors of the population became economically redundant and social tensions increased, inclusion became the political watchword but government programmes tended to focus on inclusion through skills training, welfare-to-work schemes and so forth, rather than improvement to the spatial conditions of the urban poor.

Neither economic inclusion nor communicative planning practice possess a model of spatial welfare which address the question of social cohesion, new cultural sensibilities and the non-economic, quality of life issues becoming increasingly important in everyday urban politics. In this sense the environment was a convenient means of providing substantive content to the deliberative procedures of communicative planning. The environment allowed the re-spatialization of the public interest in the context of a contracting welfare state, privatized social worlds and the diminished scope of neoliberal spatial management tools. Above all, as an historical product of local ecological systems and cultural practices, the environment provided an autochthonous sense of place in the deterritorializing world of globalization, which city planning could grasp with both place-defending fervour and a sense of global responsibility.

The debate on urban form

Whilst the Rio Summit gave particular importance to urban management, it implicitly raised the question of urban form. As we have already seen, much urban environmental analysis is concerned with resource systems and interactions, but the idea of sustainable cities required some further moment of synthesis. Beyond systems analysis was the challenge of spatial organization itself, and the environmental perspective had an invigorating influence on thinking about urban form. In effect, the extension of environmental concern to the city brought with it a tantalising new question for both environmentalists and urbanists: given the environmental crisis facing humanity as a whole, and the high urban content of that crisis, was there a settlement pattern and built form somehow intrinsically suited to fostering a sustainable future? Could a particular urban spatial configuration be devised with spatial properties sufficient in themselves to overcome the substantial economic, social, political and technological obstacles to urban sustainability? In short, sustainability opened up the possibility for the new idealization of urban form on environmental grounds.

In response to these questions, one of the opening salvos came from an unlikely source. The Commission of the European Communities' (1990) *Green Paper on the Urban Environment* proposed and extolled the virtues of the compact city as the ideal type, for Europe at least. The report strongly advocated the functionally-mixed, high-density, compact city. The compact city, it was argued, revives 'density and variety, the efficient time and energy-saving combination of social and economic functions; the chance to restore the rich architecture inherited from the past' (CEC, 1990: 19). The compact city was held to offer less natural resource consumption, the maximization of spatial and architectural capital in the restructuring of urban economies and a higher quality of life: the perfect city form for environmental sustainability, economic growth and the cultural enrichment of citizens.

The compact city idea provoked considerable reaction, given both its strong position and the institutional authority of its authorship. In the UK, the British section of the Regional Science Association held an international conference shortly afterwards, in 1991, on 'Sustainable Development and Urban Form', from which the influential book of the same title was published under the editorial guidance of Michael Breheny (1992). Breheny himself, circumspect of 'an elegant, if rather poorly substantiated vision of the future of European cities', set out to systematically examine the inconsistencies of the compact city idea. Given the impact of Breheny's work, it is worth briefly setting out the main points of his environmental critique of the compact city proposal.

First, he argued that there is no clear correlation between urban form and energy consumption. At first sight the compact city would appear to offer advantages in terms of shorter journey lengths, greater use of public transport, combined heat and power systems and reduced energy consumption in buildings. However, empirical research and theoretical modelling suggested that urban form was just one factor determining urban energy consumption and that fuel taxes, lifestyle, non-work travel behaviour and regional settlement patterns, amongst other factors, were sufficiently important to make the estimation of the direct effect of urban form on energy consumption a very hazardous exercise. Second, he questioned the view that compact cities provide a better quality of life and the feasibility of accommodating growth within existing urban areas. Demographic trends (especially in the UK) showing a flight from the city to the suburbs and small settlements, he argued, provided objective evidence of people's preferences and perceptions of a higher quality of life in dispersed, low-density settlements with less congestion, pollution and environmental degradation. Third, Breheny pointed out the threat posed by the high-density compact city to the accommodation of nature through the squeezing out of urban green space. The considerable environmental advantages of the 'green city' could be sacrificed through densification, although he acknowledged that detailed urban design could to some extent resolve this problem. Fourth, he argued that the compact city failed to take into account the opportunities for spatial decentralization afforded by communications and information technologies. Decentralized organizations and flexible home working could bring many environmental advantages, especially through reduced travel, and new communications opportunities could render redundant the city as a medium of personal contact and innovation. Fifth, he suggested that the compact city could restrict the exploitation of renewable energy resources such as wind and solar power and, finally, he warned of the possible negative effects of urban concentration for rural economic development in depressed areas.

Much subsequent debate has followed the themes outlined above (European Foundation for the Improvement of Living and Working Conditions, 1996; Jenks *et al.* 1996; Williams *et al.*, 2000; Jenks and Burgess, 2000; Williams, 2002) and despite Breheny's warning that a more careful examination of the compact city idea was needed before committing European urban policy in this direction, compactness became an established policy goal with worldwide influence. However, in addition to the inconclusive nature of the academic debate, several factors have been important in debilitating its overall position. The assumption of closed cities became increasingly problematic as spatial organization became more and more regional in scale, and indeed the specifically environmental focus on urban form has been extended to regional settlement patterns. Furthermore, the compact city idea was attacked as embodying a nostalgic vision of cities and their spatial communities, capturing but a small part of postmodern spatial form. Large swathes of urban theorizing and new urban space such as edge cities, urban 'themeing' and simulacra, segregation and gating, global city networks, and so forth (Hayward and McGlynn, 1993; Soja, 2000) breached the conceptual boundaries of the compact city idea and the growing number of large cities and huge metropolis in the developing world seem outside its practical reach.

Perhaps the most important outcome of the compact city idea has not been its impact on spatial organization or environmental significance but, rather, its contribution to the rediscovery of urban design after the demise of modernism and the disorder of deregulated urban development in the 1980s. Compactness can be achieved relatively easily in urban fragments such as city centres and residential neighbourhoods, and its environmental component connected fortuitously with emerging urban design proposals around 'responsive environments' (Bentley *et al.*, 1985), urban villages (Neal, 2003) and the New Urbanism in the USA. Above all, compactness could be usefully applied to rationalize the spatial restructuring of urban economies through city centre renewal, business and arts districts, inner area gentrification, leisure and cultural quarters, and so forth. This now staple recipe of urban revivalism facilitated the revalorization and circulation of capital through the property market and is now a key element in the definition of urban quality and competitiveness (Hall and Hubbard, 1998; Jessop, 1997; Swyngedouw *et al.*, 2002; Griffiths, 1998). In retrospect, the genius of European Commission policy was to foresee the fragmentarily compact, green and clean city as the key to urban economic futures.

The technical focus

All urban environmental agendas, whatever their philosophic and conceptual approach, seek to re-establish a balance with natural systems through modification of the exchange of material and energy undertaken in urban systems. The management of urban relations with natural resource systems leads to a quite specific technical focus on cities and a new functionalist approach to urban analysis founded on the laws of ecology and environmental engineering. The technical focus is fragmentary, isolating urban components and subjecting them to technological improvement and environmental performance standards. It embraces a wide range of initiatives which can be outlined schematically in the following terms.

Resource use efficiency

The environmental perspective emphasizes the city as a major consumer of energy and materials, especially fossil fuels and water. Two major sets of concerns are highlighted: energy conversion (sources, technology) and energy use (conservation, cost-effectiveness). Energy consumption gives especial significance to industry and eco-efficient production but also throws attention onto the technology of the production of the built environment and urban infrastructure. Early concern for integrated energy policies, sensitive to local energy resources and demand at the city-region level, has been undermined by the privatization and transnationalization of energy companies. Much more effort has gone into energy conservation and efficiency, introducing some quite concrete items on the urban agenda, such as a renewed interest in combined heat and power/district heating (in colder climates), energy conservation in building technology, construction and layout, energy audits in organizations and technological efficiency in industry and business. More recently, increasing demand and climate instability has heightened political attention on water resource management, which is more sensitive to local resource availability.

Pollution control and waste management

As centres of resource consumption, cities are seen as having a spatially dispropor-tionate contaminating capacity, the effects of which extend beyond city boundaries and relate directly to global problems such as habitat destruction and climate change. From an intra-urban perspective, pollution has become complex and sinister. Despite the reduction of some common pollutants in cities of developed countries, the array of chemicals, volatile organic compounds and toxic metals in air, water and soil systems has increased, and with it the production of 'secondary pollution' through synergism or pollutants acting in combination, whose patho-genic potential has extended in macabre fashion to the neurological and the carcinogenic. On the other hand, the sheer volume of non-hazardous waste creates considerable practical problems of disposal, and a relatively unpolluted, safe, noise-free, aesthetically attractive physical environment has become a scarce commodity, closely associated with current conceptions concerning a desirable quality of life. The containment, abatement or avoidance of pollution is now an explicitly social aspiration behind the urban agenda of integrated pollution control, monitor-ing, regulation and charges (the polluter pays principle), environmental impact assessment, and the recovery, re-use and recycling of waste.

Transport

As a major consumer of fossil fuels and source of pollution, transport tends to be given especial attention in urban environmental agendas. Improved efficiency and emission standards of the internal combustion engine have yet to be superceded by new technology, and therefore improved public transport, traffic management and integrated land use and transport planning have become important items. Trams and light railways have enjoyed a recent boom in Europe, arbitrary restrictions (by number plates, for example) on private car use are now common, and the intro-duction of road charging as in central London is likely to spread significantly, either as part of traffic management policy or through the privatization of road provision. However, the directing of increased mobility to public transport usually requires massive public commitment and investment, which run contrary to current neolib-eral policy. Urban environmental agendas also support a revival of walking and cycling as energy-efficient and healthy means of transport. Both require appropriate infrastructure provision and depend upon a spatial organization that facilitates concentrated, mixed-use development.

Disasters

Environmental disasters continue to cause enormous destruction of life and prop-erty and dramatize the functional imbalance of cities with nature. Both natural phenomena such as earthquakes and manufactured tragedies such as Chernobyl inspired preventative measures and new forms of urban regulation such as risk-mapping and anti-seismic building controls, whilst extreme weather events possibly associated with climate change are a growing concern. Cities of the developing world are commonly affected by smaller scale mudslides, floods and other hazards often not recorded in global statistics. For a long time considered to be a problem of the poor sectors of cities in developing countries, disasters are now sweeping across the cityscape of the developed world. A recent government report in the UK, for example, predicts that four million homes are now under threat of flooding.

Nature

Environmentalism attacks the 'monoculture' of cities as contrary to the biological richness and diversity of natural habitats, and argues for the need to balance town and country, manage nature within cities and build them in harmony with the nature's laws. The philosophic and aesthetic appreciation of the virtues of nature in cities has, of course, been an important strand of urbanism since industrialization. However, current concern for the greening of cities is distinctive in its scientific, ecological rationality and nature is now valued for its functional properties. Vegetation, for example, absorbs pollution (especially carbon dioxide, the major greenhouse gas) and filters particulates, regulates water run-off and ameliorates flood and landslide risk, reduces noise levels by acting as a barrier, improves micro-climates through the modification of solar radiation, wind speed and humidity, and helps minimize energy loss in buildings by providing protective cover, and alleviates stress. This urban ecology – based on augmenting urban biomass and biodiversity – leads to the transformation of cityscapes through the landscaping of available open areas, external building veneers such as roof gardens, wall-creepers and hanging plants, and a new generation of open spaces such as community forests, urban farms and wildlife parks, protected woodlands and green corridors.

There are two points we wish to make about this now familiar technical focus. First, the notion of alternative technology has disappeared as the technological challenge has been absorbed by conventional technology companies and introduced into orthodox lifestyles. The counter-cultural connotations of early alternative tech-nology proposals have been reduced to a question of costs and accounting, both for business and households, and for the urban poor this at best means 'low' as opposed to 'alternative' technology. Second, such a shift is an integral part of ecological modernization or centralized high-technology solutions to environ-mental problems. Since technology carries with it cultural values and assumptions, then the domination of the technological agenda by corporate capitalism was an essential requirement for achieving general social acquiescence to orthodox sustain-able development policy. The technological response of ecological modernization is now a routine part of everyday life, impregnated into everything from changing a light bulb to travelling to work. In other words, it is inseparable from the social agenda of urban environmentalism, which we go on to examine in the following section.

The social agenda

The social agenda of urban environmentalism emerged with force during the 1990s as debates on technical issues became absorbed by professional institutions and specialized regulatory authorities, the latter often with a strong international component such as European Union regulations, ISO 14000 or World Bank funding conditions. Outside these institutional confines, sustainable city policy needed to show itself as a credible development project, capable of bringing demonstrable improvements to urban living conditions. Even when a statistical account of environmental improvements could be presented – a reduction in carbon dioxide emissions or improved sewerage treatment rates, say – this sort of information cannot, of itself, be readily interpreted by citizens as constituting an improvement to their personal lives. Urban sustainability policy therefore had to construct meanings of welfare around the environment through its association with ideas around the quality of city living, since what may be self-evident achievements for scientific environmentalism have to be transformed into social realities by city authorities.

Equity, health and environmental justice

In general terms, sustainability policy addresses issues of equity rather than equality, and in two major ways: intergenerational equity with regard to stewardship of the earth for future generations, and intragenerational equity in the sense of conceiving poverty as a contributory cause of environmental problems and the overcoming of poverty as an aim and condition of sustainable development. The problematization of intergenerational equity is perhaps the main contribution of environmentalism to contemporary thinking on development. It is certainly the major preoccupation of most environmental movements with their ecological life-world philosophy and long-term perspective, and it underwrites their activism in everyday politics. It may also be a deep concern for many people aware of the potential implications of ecological change but distrustful of government to do much about it and unsure of their own power of agency (Macnaghten and Urry, 1998), but beyond its formal and ethical attraction it is difficult to estimate its real import with regard to everyday practice. We will therefore examine more closely the idea of intragenerational equity and the question of urban equality.

Since it is held that both wealth ('over-consumption') and poverty ('under-consumption') contribute to environmental degradation, then the reduction of inequality is an implicit objective of sustainable development. The sometimes intense polemic surrounding the poverty–environment relation (Bryant, 1997) is a spurious one given that society produces wealth and poverty simultaneously and dialectically. Whether one or the other is more environmentally destructive is logically inconsequential, although ideologically potent. For this reason, international urban sustainable development policy focuses one-sidedly (and unsuccessfully) on poverty reduction and argues that improving the environment of the poor increases their 'equity' or life chances (Hardoy et al., 2001; UN-Habitat, 2004). On the other hand, the very idea of over-consumption is anathematic to contemporary neoliberal ideology and consumer culture, and tends to be denied in practice by the illusionary effects of improved product performance and the comforting sensation of ecologically responsible behaviour in the act of consumption. At the margins of economic theory there seems to be growing political support and practical accept-ance of the idea that the wealthy must make 'sacrifices' (Nijkamp and Perrels, 1994) by paying the environmental costs of high consumption, either through the price mechanism or voluntary concessions made in the public interest: restrictions on car use, limits to freedom of action in conservation areas, the acceptance of expensive house insulation standards, and so forth.

Many environmentalists would probably agree with Blowers (2000: 106) when he suggests that 'greater equality [is] an essential precondition for the social cohesion necessary to secure a transformation towards sustainability'. However, by marginalizing the question of wealth redistribution, ecological modernization requires that such social cohesion be sought at the symbolic level. The under-consuming poor, it is held, should be helped towards better basic collective consumption of housing and services, in partnership with city authorities but without any significant transformation of urban social relations. In practice, this means focusing on areas of urban poverty in inner cities, outlying estates or shanty towns. However, as social programmes tend to be cut back or fine-focused on extreme poverty under neoliberal policy, the potential scope of environmental equality is continually thwarted by the production of new poor citizens and insufficient public investment.

Improved patterns of consumption as a whole are held to be a key factor in improving the healthiness of the urban environment. Here the environment typi-cally bifurcates into the 'green' and 'brown' agendas: aesthetic consumption and global ecological concerns for the wealthy, improved basic sanitation and an intra-urban horizon for the poor. The danger with this sort of policy perspective is not only that the holistic vision of environmentalism gets split in two, but also that environmental management practices aggravate urban inequalities. Basic sanitation and minimum environmental health become the maximum goal for the urban poor whilst the wealthy pursue unlimited aspirations of spatial amenity, with neither, in themselves, guaranteeing long-term environmental sustainability. In general, the poor inhabit the worst housing and most disaster-prone areas of cities, thereby adding intensified health risks to their already precarious conditions of existence, aggravated by the resurgence of poverty-related diseases such as tuberculosis, malaria and dengue fever and the widespread and debilitating effects of respiratory and diarrhoeic illnesses, all connected with poor physical environments. However, physical environments are social products, and the fact that a green/brown agenda can exist at all is facilitated by growing urban segregation and spatially discrete social activity patterns, especially in the cities of developing countries.

It has been argued that the notion of environmental justice is an appropriate means of reconciling the 'green' and 'brown' agendas (McGranahan *et al.*, 2001). Rejecting the tendency to blame poverty, and by implication the poor, for environ-mental degradation, it is argued that neither does environmental sustainability ensure poverty reduction nor does poverty reduction ensure environmental sustain-ability; 'sustainability is not enough' (Marcuse, 1998) and only the pursuit of equity and social justice can provide any meaningful relationship between the two.

However, the technification of health issues through environmental engineering and the subterranean nature of much of its infrastructure have a depoliticizing effect. The issue of social justice tends to be 'managed out' of basic sanitation and public health issues. Within the general trend of the privatization of health, re-politicization seems to require a visible spatial expression, for example in the case of floods and landslides in shanty towns, or to be associated with social discrimination as in the case of toxic waste disposal sites in ethnic areas of US cities.

Local democracy and urban governance

Delivering sustainable urban development has special implications for local government and spatial management practices. It is held that the integrative character of environmental solutions necessarily requires the participation of all social actors and individuals in bringing about holistic change to urban development patterns. Therefore, local authorities cannot in themselves bring about sustainability but must work with business and civil society to encourage and coordinate effective change across the whole spectrum of economic and social activity. Despite conflicting interests in the short-term, the environment is seen to be an area of collective concern where social agreement is both possible and necessary. 'Good governance' will provide the means to reach agreement on how to meet multiple goals through technically competent urban management and participatory local democracy.

These sorts of arguments are supported by environmental political theory which emphasizes the role of the environment in the legitimation of government. If environmental protection is the principal challenge for economic growth and a major demand of urban populations, then government must respond effectively to environmental issues from both an economic and social standpoint. The ensuing 'green state', it is argued, is a logical 'third transformation' following the liberal state of early bourgeois capitalism and the welfare state arising from the integration of working class interests. The precise dynamics of this 'strong ecological modernization' are seen as dependent upon well-organized environmental movements, an active environmental 'sub-politics' (Beck, 1996) and the legitimation imperatives of particular state organization, as explored by Dryzek *et al.* (2002) in relation to the political trajectories of environmentalism in Norway, Germany, the USA and the UK. Other commentators place greater emphasis on the regulatory function required from the state in the sense of the formalization of the norms of sustainability, and argue that strong and effective regulation will emerge only in the latter stages of 'ecosocialization' as a response to the intransigent demand of a social-environmental coalition of political forces (Low, 2002). This more optimistic vision of agency and the capacity of 'a myriad of small-scale, localized grass roots movements, intellectual foci, and organizations outside the main hierarchy of industry and politics' is also used to argue against the need to change urban form and instead insist on a mass of micro-level reforms in urban management practices. Finally, it is claimed that sustainable development is a learning process and that participatory democracy has the function of bringing together generalized (scientific) knowledge, professional expertise and local context-bound understandings of the environment as the condition of greater understanding and practical advancement, as explored by Edén *et al.* (2000) in relation to the Swedish context where state policy and regulation is already well advanced.

The gigantic assumptions concerning the inherently democratic character of environmentalism have been challenged through empirical urban research. In part, such claims are based on the ability of the environment to reactivate local democracy through widespread participation, but there is much evidence and widespread concern over the generally low level of citizen participation in urban sustainability programmes, whether Local Agenda 21s or more specific environmental projects. At stake is much more than a quantitative calculation, important though numbers are in any democratic process. Democracy is also about the distribution of power and social justice, and environmentalism can have negative effects on both accounts. Noting the small-scale and barely significant nature of sustainability achievements in the UK, Wild (2000), for example, reports on the credit union, local exchange trading schemes and community farming and transport projects undertaken as part of urban regeneration in deprived communities. Not only did he find that the scale of the projects was small and participation low but also that such projects have a subordinate role in relation to mainstream planning and economic regeneration, tend to be voluntary rather than enforceable through regulation and that 'very often, the environment is only protected, and excluded groups only included, when this does not impinge on the ability to attract inward investment'. Wild illustrates how the situation of disadvantaged groups may actually be worsened by urban environmental management. Citing the case of Sheffield, UK, he shows how the urban poor are substantially excluded from the economic benefits of business-led urban regeneration (in terms of jobs, local community facilities, property market performance) whilst at the same time bearing a disproportionate share of the externalities (increased traffic and pollution and associated health problems).

Neither does participatory democracy necessarily guarantee better sustainability policy, as Hajer and Kesselring (1999) illustrate in their study of transport policy in Munich. Their analysis reveals the existence of a double circuit of distinctive discourses and two-tier policy deliberation: the power politics of entrepreneurial governance (think tanks, round tables, strategic partnerships) in league with corporate interests existing alongside more cooperative management régimes aimed at difficult, deliberative consensus building with civic organizations. Hajer and Kesselring found that the former, though less democratic, can come up with more sustainable solutions, at least in the ecological modernization sense, in this case through a BMW-led coalition promoting innovative technological solutions for urban mobility which promise to be more effective than conventional 'green' traffic management policies. Ecological modernization of this sort avoids the trade-off dilemmas typically faced by local authorities when having to choose between economic development and environmental protection. In restructuring urban economies with high unemployment rates, city authorities are extremely vulnerable to adverse local media presentation of the jobs-versus-environment equation and tend to willingly forego environmental protection in favour of investment and to sacrifice long-term learning for short-term fiscal advantage (Wild, 2000). When 'smart growth' can apparently eliminate such trade-offs, the balance of sustainable urban development policy tilts decisively towards big business and unaccountable forms of government; high-technology innovation, not democratic deliberation and public regulation, become the key words. In this kind of process little account is taken of the limits of eco-efficiency and smart growth in the face of increasing cumulative consumption, edified in cities through expanding infrastructure and building stock, as observed in the case of Norway and Denmark by Høyer and Næss (2001).

This in turn raises the question, as Hajer and Kesselring (1999) point out, of who sets the sustainability agenda, its policy-frames and cognitive maps. The competitive dynamics of private enterprise and technological development outpace slow democratic deliberation and consensus building, elbowing out idealized models of collective knowledge construction and circulation. Ecological modernization is increasingly effective in its own terms but far from democratic in its operation and political consequences. 'Its own terms' consist not only of technological solutions to discrete environmental problems but also of maintaining the balance of power in terms of effective knowledge and policy strategies. The outcome of executive-type public–private coalitions can pervert long-term sustainability as well as have hugely undemocratic (disempowering) effects through subtle and persuasive means, as Lockie (2000) demonstrates through an analysis of the Australian Landcare programme. Here, ecological modernization within a neoliberal policy framework not only integrates people into business-led technological solutions and their policy settings, but also 'shapes the boundaries and objects of legitimate government', or the distribution of responsibilities between government and citizens. This distribution is mediated by the demands of capital accumulation rather than sustainable practices *per se*, in a political economy and not political ecology of the environment. A similar conclusion is reached by Bauriedl and Wissen (2002) in their study of competitiveness and sustainability policies of the Hamburg region in Germany, where they show how nature is 'constructed' as a function of economic development in the narrative of ecological modernization, in the sense of a competitiveness factor for attracting investment and providing 'quality of life' for employees whilst simultaneously absorbing some of the contradictions of post-Fordist space by constituting a new field of political contestation.

In cities of the developing world, the question of governance receives particular attention in the sense that management capacity-building is required to improve service delivery and develop partnerships with the private sector and communities, whilst the strengthening of local democratic processes is held to be necessary to reduce corruption and improve efficiency. For the urban poor, environmental improvements are seen as a direct way of integrating the informal sector into the city's physical development, at the same time providing them with leverage with which to empower themselves in political terms. However, as Dragsbaek Achmidt (1998) has observed with regard to South-East Asia, the potential for this kind of process is often limited by a weak 'social contract', extreme social inequality and conflict, with the combination of weak nation-states and dependent local authorities unable to respond to the social and environmental pressures arising from a globalized economy and neoliberal urban policy frameworks.

To sum up, we can say that evidence to suggest that the environment strengthens democracy is flimsy at best, and the critical study of urban environmentalism suggests that, in itself, environmental protection neither requires nor promotes participation. The conceptual argument in favour rests on the assumption that the environment is a collective concern where consensus can be achieved and best managed at the city level. Such a proposition ignores the power politics behind the urban environment and ecological modernization. Power interests may occasionally reveal themselves in the face of public protest around a particular urban project such as a motorway or industrial plant, but normally they operate in more subtle and influential ways: through policy alliances, the unequal distribution of knowledge/meanings and differential spatial regulation. The formal search for social consensus around the environment covers over these discriminatory procedures and underlying conflicts of interests. If, as Flyvbjerg (2001) argues, democracy is as much about conflict as consensus, then the bland politics of much urban environmentalism is not a healthy indicator of anything other than the success of the environment as a state legitimation strategy.

Community

The notion of community-building through the environment was only formally introduced some time after the Rio Summit. It had been an implicit concern of both conservative and counter-cultural environmentalism from the 1960s but initially failed to connect with much early sustainable development theory. However, the Rio Summit emphasis on locality and incremental practice required city governments to work with local communities for effective sustainability delivery. Urban environmental agendas have emphasized communities both in a functional sense and in terms of the duties and responsibilities forming the basis of a renewed sense of local commitment and collective interest. For successful policy implementation, basic cooperation is required at the community level for the mundane but environmentally fundamental programmes such as garbage collection and waste recycling, energy and water savings, local traffic management schemes, open space upkeep, urban farming, collective garden composting, the control of crime, and so forth.

However, as public policy it may be argued that community-building through the environment only flourished in response to political preoccupation around the decline of local communities in general and the increasingly severe problem of how to administer disadvantaged social groups concentrated in pockets of urban decay. As Goodchild and Cole (2001) remind us, social balance through community building has always been the preserve of deprived neighbourhood policy. On the other hand, centralized market-based approaches to environmental citizenship only encourage individualism and the continuing decline of community in more affluent areas (Smith *et al.*, 2000). As a result, during the 1990s an abstract environmental aspiration – community-building – became a political convenience of considerable importance in the administration of space.

The notion of stable residential communities has long been a staple part of spatial administration. The assumption of people sharing a life-space, growing up and growing old together within a web of shared class or cultural interests, kinship relations and friendship bonds was for a long time both a social reality and the conceptual basis of urban planning and management. However, economic restructuring began to demolish the remaining foundations of community from the 1980s onwards. Labour markets and the consumption of services such as education and health became delocalized, alongside a dramatic increase in spatial (but not social) mobility, as the conduct of social lives was radically decoupled from propinquity or a sense of neighbourliness. This general pattern has, of course, marked variations, with the spatially-liberated globalized elites at one extreme and the spatially-tied 'underclass' at the other. However, the evaporating reality of spatial communities became a serious problem, not only in the sense of an increasingly spurious assumption of spatial management or a disappearing intermediate social unit, but more to the degree that belonging to a community implied social responsibility and moral obligation: its cultural codes both minimized the need for and facilitated the implementation of other sorts of social regulation such as local laws and policing.

Environmentally inspired community management also fits into attempts to re-establish spatial communities through urban design and eco-neighbourhood technology. High-density mixed-use residential areas of the new urban villages variety mentioned earlier are strongly promoted on the basis of their propensity for encouraging a sense of community. Promotional literature for one such proposal, for example, states that: 'The Forest Village would be a balanced community for people of all ages and incomes, where people can live, work and enjoy a vibrant community life, the majority without the need to commute and where everyone could feel a sense of belonging' (cited in Barton, 1998: 162). Widely criticized for their conservative nostalgia for pre-modern sociality, urban villages have much in common with gated residential areas and spatial segregation through expanding house price differentials and joint property regulations.

Nevertheless, underlying urban environmentalism is a deep-seated conviction that community living is both desirable and possible, and that the environment is a means to reinvigorate it in the face of contemporary urban challenges. However, experience to date has been disappointing and studies suggest that there are considerable obstacles. Certainly community involvement in Local Agenda 21 programmes has been disappointing, along with fatigue around complex community partnership projects in general. The explanation for this situation surely lies not in the formal weaknesses of urban environmental proposals but in the social and political realities of contemporary spatial groups. The Danish case, where considerable efforts have been made to introduce sustainable urban development through community action, provides a useful illustration. Gram-Hanssen (2000) describes the basic elements of what is referred to as 'the local ecological dream' of closed resource cycles and local communities as promoted by the Danish authorities, wherein the recycling of energy, waste, water and other materials is both an ecological end in itself and a means to re-establish and develop local community life. Using Tönnies' distinction and tensions between *gemeinschaft* (traditional community) and *gesellschaft* (community in modern society) as a means of describing the type of social relations that exist within housing areas, three 'ordinary neighbourhoods' (as opposed to eco-villages or environmentally committed housing areas) of different socio-economic status were studied. It was found that what actually happens on the ground, in terms of both the environmental protection and community building, is largely independent of the rationality underpinning such programmes. Wealthy home-ownership areas may adopt ecological programmes without necessarily developing stronger community ties, the strong collective management organization of housing association areas may release inhabitants from any sense of individual responsibility and mutual cooperation, and in poorer public housing areas environmental programmes managed from above may temporarily coalesce collective interests without producing any significant environmental change or community consolidation. The considerable gap between the ecological dream and real outcomes appeared to depend more on the type and trajectory of the existing community relations, the condition of labour and housing markets, the political relationships with the local authority and overall municipal policy. In all cases, a critical factor in the success or otherwise of such programmes was the role of environmental activists rather than the community itself.

Community building through Local Agenda 21s and environmental issues generally would seem to be subject to these same indeterminations. The environment may temporarily both unite or divide local opinion, without necessarily modifying local social relations, depending on particular contexts and conjunctures. The existence of environmental threats to local health and amenity due to hazardous wastes, contamination or flood and landslide risk may provoke community solidarity; equally, environmental measures which imply increased or differential costs in service provision or highlight sub-group interests around questions of noise and disturbance may divide communities. In socially and ethnically diverse areas, these sorts of problems are easily exacerbated through different political interests in and cultural understandings of the environment. Whatever the case, urban environmental management appears to offer little guarantee of community-building in any significant sense.

Get a life: lifestyles and livelihoods

Whilst global environmental issues are long-term and only detectable through scientific knowledge, cities allow them to become immediate and tangible. Hugely complex phenomena such as climate change and loss of biodiversity can be simplified and symbolized in concrete things such as automobile emissions, domestic waste management and urban green space. This urbanization of sustainability not only localizes global issues but also personalizes them in shared geographical life-spaces, with the effect that sustainability becomes an experiential concern. Whilst city authorities assume overall responsibility for making the functioning of cities more environment-friendly and appeal to diverse 'communities' to assist them in undertaking policy implementation, a complementary argument is that individuals must also assume their part and be willing to adjust their ways of urban living. The combined effect of public management and personal change will, it is argued, be better quality cities and an improved quality of urban life.

To a certain extent this is an obvious proposition. A less polluted environment means fewer health risks, litter and noise offend most people, the majority of people enjoy a park where they can play with the kids or walk the dog, and so forth. However, urban environmentalism has raised the profile of these sorts of everyday issues to the level of definitive criteria of the quality of the urban experience. Given that this experience is never an isolated one, it forces individuals to think not only about their own behaviour but also about the conduct of others. The quality of urban life is always dependent on the collective experience of cities, as the transport issue clearly illustrates. The ever-greater use of the private automobile results in congestion, contamination, accidents, wasted time and psychological stress, until the point is reached when individual freedoms and unrestricted personal consumption have to be curtailed in order to restore some sort of qualitative balance back into everyday life.

The cumulative effect of this sort of phenomenon is commonly described nowadays as the 'livability' of a city, or the degree of social, physiological and psychological well-being afforded by urban environments. Healthier social and physical environments arising from better sanitation, lower levels of pollution, diminished stress and improved lifestyle opportunities, as well as their conduciveness to social contact, diversity of experience, access to and quality of services, facilities and green space, place identity, safety, security and health are now the topic of city indicators and league tables, the latter being a new form of evaluation of cities measured according to the requirements of the mobile elites.

The specifically environmental contribution to the idea of the quality of life incorporates everything from abstract notions of 'living with nature' to concrete concerns over the personal health and safety implications of immediate surroundings. It represents a fundamental shift from quantitative to qualitative evaluation, from material to aesthetic values, from work time to free time, from self-interest to seeing individual well-being as dependent upon that of others. However, it presupposes personal responsibility in individual decision-making, the ability of individuals to act both rationally and ethically, and the availability of choice as to the conduct of everyday lives. A 'green' urban lifestyle is held to be an option which people can freely take up with regard to home energy consumption, travel modes and patterns, the value placed on and use of open space, the degree of tolerance of noise and anti-social behaviour, and so forth.

This environmental slant on lifestyle is, of course, part of a general condition of individualized contemporary society. The colour magazine version of lifestyle is about expanded and diversified consumption; the environmental edition invites restrictions and technological alternatives, although both are underwritten by the notion of choice and individual responsibility (for personal identity or ecological survival) and the permanent need to consult and measure one's personal perform-ance (through fashion or ecological footprint). Lifestyle, though founded on individual choice, is no longer an option but now an inescapable social requirement, and an environmental lifestyle, of itself or tagged on to others, ceases to be an area of free expression and becomes transformed, like all the others, into an obligatory framework of rules and performance requirements. An environmentally responsible lifestyle, however, is perhaps the only one which dares to claim universal validity.

Lifestyle, of course, is for the wealthy; for the poor, there are only livelihoods. Lifestyle presupposes a level of disposable income sufficient for the exercise of consumer choice, which the urban poor are categorically denied. Faced with grinding routines, the poor are offered only strategies of survival and the struggle for integration into the income-generating activities of cities. Here, government assistance is required to provide the minimum conditions of habitation, health and life-chance improvement.

> Almost a billion people – or over 40% of the urban population in the developing world – live in slum settlements; and if current trends continue, the number will more than double over the next 30 years. [. . .] What is often forgotten is that when water and sanitation services are planned and provided as an integral part of settlements development, it contributes to livelihood development through income and employment generation and can help address overarching poverty and environmental issues more effectively. With experience and understanding also comes the recognition that urban and shelter finance mechanisms are essential to poverty reduction and are best achieved through collaborative partnerships.
>
> (Anna Kajumulo Tibaijuka, Executive Director, UN-Habitat, 2004)

Livelihood focuses on the physical environment and its enabling capacity whilst ignoring the economic conditions which force people into poverty as a social condition, and the neoliberal fiscal and public policies which keep them there, such as regressive taxation, public spending restrictions, the removal of subsidies, the privatization and marketization of public services, the flexibilization of labour markets and lower basic wages, and so forth. Whilst centring on the poor in the cities of the South, the same processes apply equally to the poor in the cities in the developed world.

Conclusions

In this chapter we have outlined the contours of the substantive content of urban environmental agendas whilst also signalling the conditions within which they arose and the arguments typically employed to rationalize them. In this immensely wide field, it is clear that urban environmental agendas have taken up and spatialized the argumentative rationale and thematic evolution of orthodox global sustainable development policy. However, behind these formal challenges being addressed by spatial planning and management, it is equally clear that a fuller understanding is incomplete without reference to other changes affecting cities over the same period, in particular globalization and the social effects and political implications of the restructuring of urban economies.

Urban environmental agendas have not, in themselves, transformed cities or city-regions but they have undoubtedly had a significant impact. On the basis of a city's relation to natural resource systems, new avenues have been explored which might contribute to making cities not only more environment-friendly but also healthier, more democratic and offering a better quality of life. We have focused on general proposals – environmental ways of conceptualizing, representing, organizing, building and managing cities – rather than concrete achievements. Indeed, although significant urban innovations have been realized, sustainability in general is characterized by the persuasiveness of its propositions rather than its demonstrability as a reality. The sustainable city, like sustainable development in general, is a challenge and prospect, something to be sought for and worked towards. Most practical attention is directed towards measuring concrete progress, but the fact that the sustainable city has to be conceived before being achieved makes the examination of general proposals that much more important.

In these conclusions we want to highlight two outstanding conceptual devices of urban environmental agendas. First, an ecological functionalism, applied with technical rigour or metaphoric allusion, underpins everything from infrastructure provision to quality of life issues. Second, this analytic and allusive device requires the reification of the city or its objectification as a 'thing' rather than the constellation of changing flows and relations. Reification solidifies the economic transformation of cities and urban difference by focusing on the city as a 'living organism' or 'energy transforming machine', with the consequence that social processes are not seen primarily as producing the city but rather as secondary phenomena which need to be adjusted in function of the object to be transformed or made sustainable. With reification, transformation of the city rather than the social processes which produce it, holds the key to better urban futures. The Berlin conference is an outstanding example, with the city presented as an island of social harmony and human fulfilment amid the swirling currents of global geopolitics.

Urban environmental agendas in general are based on this conceptual assumption, without doubt the cause of much technical frustration and professional disappointment. Some environmental improvements have undoubtedly taken place in a socially and geographically uneven way, although it is generally accepted that cities as a whole are a long way from achieving ecological sustainable forms. A much stronger impact has been on the ways of thinking about cities and how this relates to social change. The normative and action-oriented nature of urban environmental agendas conceives such change in terms of the exigencies of a scientific description of ecological problems, from which social transformation must be forthcoming. Social change is held to be the requirement of the consequences of ecological deterioration. We have tried to demonstrate provisionally, albeit through somewhat limited empirical evidence and a still restricted range of critical studies, that this simplistic assumption is misleading and that the key to the understanding of urban environmental agendas lies in the relation between the rationale of scientific description and social logic of spatial transformation.

The first step in this direction is the recognition that sustainability proposals in general, and urban environmental agendas in particular, have been formulated within a period of the restructuring of capital and the predominance of the logic of the market. This has been at once a global phenomenon and an urban one, with cities acquiring heightened status as spatial organizers of the global economy. In this post-industrial urbanization driven by the financial services, communications and culture industries requiring new spatial facilities, novel qualitative criteria arose which facilitated the accommodation of ecological concerns. At the same time, privatization and neoliberal government strategies have opened open the way for a technological response in the interests and image of corporate capitalist interests.

However, within these general dynamics of ecological modernization lie some specifically urban problems relating to the social transformation of the urban experience and the politics of city governance. Economic restructuring involved a general privatization of the social world, the redesign of state welfare, the break-up of spatial communities and individualization. The legitimacy of local government, social cohesion and place identity were all undermined by globalizing forces and neoliberal state policies and the transformation of work and welfare. Global sustainability policy already had the discursive resources to address these issues in a general way, in the rhetorical sense that economic growth would lead to social equity and depended on environmental protection. Urban environmental agendas assumed the task of giving concrete form to such rhetoric, invoking participatory local democracy, health and well-being, community building and new lifestyles within their transformative reach. Existing more as discursive constructions than urban realities, their significance should nevertheless not be underestimated. The issues raised in urban environmental agendas were not a fabrication of the sustainability debate or an abstract artifice of capital, since they evidently resonated with the changing social conditions and cultural experience of urban citizens. In the following chapter we will examine how this came about.

Chapter 3
The environment and postmodern spatial consciousness

Sustainable cities and postmodern lives

The most outwardly striking characteristic of the environmental agendas examined in the previous chapter concerns the institutionalization, technification and normalizing effects of urban spatial management. Urban environmental management is presented as the rational organization of collective effort to attune urban spatial patterns and organizational and individual behaviour to the exigency of reversing or at least containing a common, global threat. However, this proposition of urban environmental management is both its strength and its weakness. It obviously has some credence among urban populations, at the same time as attempted 'rational' responses to the ecological problem constantly falter on the complexity of the overall demands of urban life and the unresolved dilemmas of environmental decision-making. In this chapter we examine some of the features of contemporary urban society and culture which at least make urban environmentalism a politically feasible project. We refer not to green politics or urban policy but, rather, the conditions under which citizens are disposed, one way or another, reluctantly or enthusiastically, to take the environment as a matter of sufficient seriousness as to merit public administration and accept its general consequences in the course of their everyday lives.

In this sense it is not ecological rationale which determines the possibilities of urban environmental agendas but the general conditions under which urban lives are lived, and here two fundamental shifts need to be taken into account. First, urban environmentalism emerged and became institutionalized within the processes of neoliberal urbanization, privatization and the tendency within representative local democracy to operate on the basis of a mixture of entrepreneurial government and partnerships. Second, and as an integral part of the first, both urban space and urban social life have become fragmented and individualized. As a consequence, the rise of urban environmentalism needs to be examined in relation to the decline of the public sphere in its modern institutionalized form, both part of and a response to the colonization of the 'public' by the 'private' and the general cultural trend of individualized society in which 'the personal is political'.

However, since cities are a shared experience, even when this experience is open and individualized, we will explore the question of how and to what extent the environment contains the elements for reconstructing a sense of the common interest, outline the cultural conditions which broadly determine whether this is a feasible project and examine some of the urban dynamics which set limits and conditions to such an enterprise. The technical agenda of urban environmentalism, in pursuing a narrow version of urban change, largely takes for granted these social and cultural considerations and assumes that citizens will, eventually, be persuaded of the importance of 'environmental rationality'. Urban environmental agendas implicitly recognize that most people are not environmentalists in the sense of scientifically informed and politically motivated citizens operating on some kind of explicitly environmental platform, yet exploit and cultivate a general citizen sensibility towards environmental problems. More than a single issue, the environment has become a frame of reference for describing people's general spatial experience. As a practical concern, the environment has not so much provoked a formal political response as permeated spatial consciousness and, therefore, come to mediate all spatial practices and the matrix of issues within which people debate their everyday spatial existence. The sorts of environmental issues which appear on a daily basis in the media – asthma epidemics and car pollution, cancer risks and the ozone layer, the security/amenity effects of the siting of nuclear power stations, the noise and night-glow of road projects, nature conservation and access to the countryside, wildlife habitat protection and biodiversity, the safety of food production methods, and so forth – have come to dominate everyday spatial politics.

The technical agenda promoted by urban authorities and the spatial consciousness of urban populations tend to converge increasingly around the notion of lifestyle. Living a sustainable lifestyle hinges on adjusting everyday habits of consumption in both private space (eating, waste management and energy consumption in the home) and public space (transportation habits, recreational activities and the use of public space in the city), and urban environmental agendas aim at facilitating such behavioural adjustments through change in urban form, building technology, infrastructure provision and management practices. Lifestyle, of course, is a highly individualized focus of urban change and partly transfers the logical contradictions and practical difficulties of making cities sustainable on to the individual citizen. Lifestyle is about people making choices in an individualized society, or more exactly, the obligation to make choices in relation to fields of personal behaviour and moral judgement which were once determined by class, tradition and local culture. In this sense sustainable city ideas need to be seen as part of a process of urban social change in which collective aims, such as making cities sustainable, are increasingly the responsibility of individuals rather than the state. However, lifestyle, for all the obsessive attention it receives in glossy magazines and Sunday newspaper supplements, is not a question of unrestricted choice on the part of individuals. It involves the seduction, awakening of desire and dedicated behaviour of the citizen-consumer but, as a question of consumption, a sustainable lifestyle is also circumscribed within income levels and social opportunity and, to a lesser extent, social group identity and cultural tradition. Only recently has lifestyle become a significant feature of social life.

The extensive shift in spatial sensibility towards the environment that has occurred over the past two decades or so, and the question of lifestyle thereby implicated, is a distinctive yet integral part of the transformation of the cultural experience of space and time in the conditions of postmodernity. If urban planning and design is the institutional organization of space and time (Harvey, 1989), then in what follows we will outline the cultural conditions which introduce the environment into individual consciousness and social practice, and so give life and legitimacy to urban environmental agendas. From this perspective, urban environmentalism acquires a specific meaning: the intellectual and/or emotional disposition to privilege, interpret or explain the dilemmas of contemporary socio-spatial experience in terms of transformed nature.

Environmentalism and postmodernity

The term 'postmodern' has lost prominence in urban studies since its heyday in the 1980s and early 1990s. What postmodern analysis tried to capture was the bewildering character of spatial change which affected cities over that period. First evident in a new style of architecture and a conscious break with the rigid geometry, semantics and social purpose of modernist buildings, the term 'postmodern' was articulated to the idea of 'post-industrial' society to describe the social patterns and cultural life of transforming cities, their economies and the politics of their management. When understood as a cultural condition we believe that postmodernity is still a useful concept to describe the conditions under which we experience time and space and the sensibilities which form around their concretization in cities, and so explore the emergence of urban environmentalism.

On the other hand, the diminished use of the term 'postmodern' is surely due to the fact that what shook our senses during the 1980s has now become part of our everyday experience and that particular aspects of that routinized experience now dominate intellectual attention under other categories: globalization, the information society and virtual reality or, at a specifically urban level, fragmentation, cultural consumption, competitiveness, place marketing, the heritage industry, and so on. In this sense postmodernity is no longer the disturbing transformation of the experience of social life, but rather the taken-for-granted, routinized and normalized parameter of our daily lives. Something similar occurred in the case of the environment. Like postmodernism, the environment was once at the forefront of intellectual debate and new ideas on what the city could and should be like, but now, integrated into urban management practices, has become a serious but rather dull dimension of policy debate, at best a field of duty and responsibility rather than an inspirational source of urban change. As we argued earlier, this integration of the environment into our daily thinking about cities is a remarkable phenomenon which has gained rather than lost significance by having become sedimented into institutional practices and social sensibilities. In trying to understand how this happened it is useful to recall the general conditions within which urban environmentalism emerged: the coming to terms with postmodernity and postmodern urban change.

For a long time, postmodernism and environmentalism were the estranged twins of contemporary social critique, containing apparent similarities within very disparate visions of the world (see, for example, Bordessa, 1993; Hannigan, 1995; Gare, 1995; Gandy; 1996). The genesis of both these fields of debate is commonly registered in the period around the late 1960s and early 1970s, but communication between them was weak and relations unsympathetic. At first sight there would appear to be an obvious explanation for such a situation. They constructed and reproduced themselves in different discursive habitats: postmodernism in the socially transformed world as reconstructed in art, philosophy and politics; environmentalism, in the world of transformed nature and the laws of ecology. Their intellectual objects and inclinations were also radically different: postmodernism filtered into every conceivable theoretical discussion of a social and cultural kind (Callinicos, 1989) whilst environmentalism penetrated all possible practical issues. Third, both contained an amplitude of perspective and inner polemics sufficient in themselves to inwardly focus attention whilst ignoring their frayed and imprecise boundaries. It is only by understanding postmodernity as a condition that the two fields of debate encounter common ground, and Huyssens' (1984) broad description of postmodernity as 'a noticeable shift in sensibility, practices and discourse formations' is particularly useful in this respect.

Bauman (1992a) describes postmodernity as modernity taking a critical look at itself. But what was modernity? Lechner (1989) argues that modernity was a slow transition from a received, divine order to a socially produced one. In this sense, it was a long and tortuous shift from a mythical to a rational-scientific explanation of the natural and social worlds, which reached its most confident expression in eighteenth century Enlightenment thought. Through the postmodern debate, three major aspects of the 'modern project' came under critical scrutiny: first, the idea of rationality, and particularly the instrumental rationality encapsulated in a scientific view of the world; second, the idea of history as a coherent linear development of civilization, with its Eurocentric assumptions; and third, and intimately related to the first two, the idea of progress founded on a confidence in the rational action of society upon itself to produce ever greater levels of freedom, welfare and self-realization. It has been pointed out that these were far from uncontested issues throughout modernity (Callinicos, 1989; Harvey, 1989). However, postmodern inquiry developed a generalized suspicion of and aversion to universal explanations of the world or meta-narratives (Lyotard, 1984), and the themes of diversity, difference and discontinuity came to dominate philosophical, political and aesthetic reflection, along with the problems posed with regard to representation, whether in the realm of truth, power or beauty.

This very schematic introduction of the themes of postmodernity, which will be taken up and developed later in relation to spatial practices, is a useful starting point for identifying some fundamental points of contact of environmentalism with postmodern critique. Environmentalism contains the idea that instrumental rationality (economic development) must take into account the laws of nature (ecology); it is founded on a recognition of difference based on the geographical-ecological diversity of life; and the environmental crisis materializes the illusion of unlimited progress. These three very postmodern premises constitute the underlying assumptions of environmentalism. However, theoretical reflection of this sort can hardly explain, in itself, the practical concern of both planning and people for the environment.

An alternative and perhaps more pertinent approach with regard to spatial practices lies in the complementary view of the postmodern as a 'structure of feeling' (Williams, 1977; Pfeil, 1988 ; Harvey, 1990). This is a perspective that has been explored more directly through the sociological examination of the 'postmodern condition' as the cultural experience of social change. A useful synthesis of this experience can be derived from an examination of the key descriptors found in socio-logical inquiry, notwithstanding considerable disagreement over the appropriate-ness of the term 'postmodern' itself. Thus Giddens (1990), for example, uses the term 'disorientation' to express 'the sense that many of us have of being caught up in a universe of events we do not fully understand, and which seems in large part outside our control' in 'high modernity'. In a similar vein, Mongardi (1992) writes of 'unease' as a distinguishing feature of the experience and reproduction of 'con-temporary modernity', due to the loss of control over reality. Lechner (1989) prefers the term 'perplexity' to describe the postmodern state of mind produced under such conditions. It evidently reflects what Bauman (1992a) terms the 'incoherence' of contemporary life. For Maffesoli (1991) postmodernity is an epigram for 'precari-ousness'. Vattimo (1988) sees in mass-mediated late modernity the 'erosion of a sense of reality' and rootlessness. Lipovetsky (1990) underlines the traumatic qualities of hedonistic society, Jameson (1984) its 'depthlessness', and so forth.

Disorientation, unease, rootlessness, incoherence, precariousness, perplexity, depthlessness and the dissatisfactions of the pleasure principle: the disconcerting litany of the sombre side of the postmodern condition. But now, into the twenty-first century, the impermanence, contingency and volatility of contemporary social life is no longer considered to be a legitimate indication of cultural disorder or cause for psychological disquiet. It has been assimilated as a normal part of everyday life and transformed into a challenge, and an increasingly individualized one, of post-modern citizenship under flexible and globalized capitalism. Theoretical explanation from a geographical perspective has highlighted the deterritorialization of economic, social and political life (O'Tuathail, 1998), uneven historical-geographical devel-opment (Harvey, 2000), global information and telecommunications networks (Castells, 1996), supra-national forms of governance (Agnew, 1996), hybridized cultures (Soja, 1996) and so forth, which describe an acceleration of the experience of time and space, an historically unprecedented bombardment of the 'here' and 'now', a perpetual change and instability in material life that blurs and erases the certainties which could develop from more fixed and stable spatio-temporal coordinates of experience.

From this transformation of social space through the annihilation of the friction of distance, along with the residues of old unease and the new dilemmas that go with it, has re-emerged a preoccupation for natural space and its stable ecological organization as an anchor for the existential certainties that space as place has always provided, and hence the formation of a particular form of spatial consciousness: an environmental one.

Ontological insecurity

Postmodern inquiry raises issues concerning our sense of being in the world. Ontological security is described by Giddens (1990: 92) as 'the confidence that most human beings have in the continuity of their self-identity and in the constancy of their surrounding social and material environments'. Our sense of being in the world is, he claims, 'an emotional, rather than cognitive, phenomenon, and it is rooted in the unconscious'. The context for Giddens' discussion is that of the sense of instability arising from the acceleration of change in contemporary society, to which he gives particular importance to the idea of abstract systems. Giddens describes these as symbolic tokens (especially money) and expert systems (of technical accomplishment) through which complex modern society operates. They are, amongst other things, 'disembedding' mechanisms which 'lift out' social activity from localized contexts. Through the lubrication of 'space-time distanciation' and all the opportunities which such a process affords, direct personal contact is substituted by impersonal systems of interaction, and hence the anonymity of modern social life. This is, of course, an old theme in sociology, which Giddens goes on to explore by pursuing the simultaneous 're-embedding' of social relations in the time-space conditions of what he prefers to describe as high modernity.

When highlighting the significance of trust, or confidence in the reliability of systems, Giddens provides an interesting insight into the postmodern condition. He argues that trust in abstract systems is the condition of large areas of security in day-to-day life which modern institutions offer as compared to the traditional world, and that the routines which are integrated with abstract systems therefore become central to ontological security in conditions of modernity; yet 'this situation also creates novel forms of psychological vulnerability, and trust in abstract systems is not psychologically rewarding in the way in which trust in persons is' (Giddens, 1990: 113).

Vulnerability and unease are, it would seem, an essential part of postmodern subjectivity, its emotional and psychological inner core. In traditional society, continues Giddens, trust was constructed on the basis of kinship, community, tradition and religion. In their place, modernity gradually imposed the rational construction of trust due to the emergence of impersonalized, abstract systems of social organization. In other words, modernity stretches trust to its limits. First, through its very complexity: in whom and in what to trust? Second, due to the human creation of the universe of possible events: uncertainty ceased to be attributable to divine providence. Third, the obligatory transfer of trust from persons to abstract systems gives it a rational rather than emotional form, thereby wrenching it from the realm of the unconscious. Fourth, if trust is concerned with the reliability of systems, the pace of change in contemporary society and the apparent contingency of events appears to render even such a concern itself somehow redundant.

However, if trust cedes to the challenges of living in a constantly changing present, the psychological residues of such conditions of existence accumulate inexorably. From this perspective, it is easy to see how the instability of postmodern sociality and the apocalyptic scenarios of ecological crisis converge hopelessly. Life itself seems to teeter on the post-prefixed ledge overhanging the end of everything. Giddens states that 'basic trust in the continuity of the world must be anchored in the simple conviction that it will continue, and this is something of which we cannot be entirely sure'. This is a basic tenet of environmentalism which is intimately connected to contemporary social experience in general. Nuclear accidents, climate change, genetic manipulation, new physiological illnesses (such as Aids) and technological pathologies (the millennium bug, for example) emerge and hang over us, to produce what Clark has called 'panic ecology':

> Mediated ecological concern may engender attempts to rationalise the human exchange with the environment (Baudrillard, 1993, 104–5). But the frenetic installation of safety measures is accompanied by an equally frantic drive to reconstitute that which is felt to be disappearing – be it endangered species, vanishing human tribes or energy resources – in what Baudrillard has earlier referred to as 'a panic-stricken production of the real and the referential' (1983: 13). On both accounts, our desperate attempts to reconcile ourselves with nature involve the rendering of the natural world into transparency, its coercion into visibility and communication.
>
> (Clark, 1997: 80)

In short, the extension of abstract systems into more and more aspects of everyday life can be seen as creating a sense of vulnerability, of which the environment is an integral part. The environment both symbolizes and materializes that sense of vulnerability, and environmentalism as a generalized attitude towards space can only be fully understood in the context of a cultural condition of ontological insecurity. At the level of spatial practices, the environmental turn of urban planning and management can, in consequence, be glimpsed in its fuller significance: not just the rational management of nature, but also its aesthetic representation in space as a comforting gesture towards ontological security and control. Urban environmentalism, it might be argued, seeks to make nature transparent and visible, in an exorcizing response to the fears of not only ecological destruction but also an unstable social order.

Risk

Ontological insecurity brings with it a heightened awareness of risk, a central element of the unsettling side of modernity. Risk may be said to be the fate of modernity, for risk is not just the presence of danger but the awareness of it, and in the very act of internalization and rationalization of danger, the negative unexpected assumes the form of human responsibility. Living with risk is the human cost of the desacralization of the world undertaken by modernity; that which had been attributed to divine providence became a mundane human affair. In Giddens' (1990: 34) terms, risk represents 'an alteration in the perception of determination and contingency, such that human imperatives, natural causes and chance reign in place of religious cosmologies'.

With the technological development of modernity, risks are not just intellectually created but also take concrete form and proliferate in everyday life. Social and physical environments are reordered in real as well as abstract terms, and in such a way that the ever-greater exercise of rational control proportionally increases the likelihood of unexpected outcomes. In modernity, risks are socially created and therefore calculable on the basis of knowledge, but never with absolute certainty. Risk is the phantom of modernity, for it means accepting, as Giddens (1990) points out, 'not only the possibility that things might go wrong, but also that this possibility cannot be eliminated'. For Giddens, high modernity has become a nightmarish concoction of low-probability, high-consequence risks such as nuclear war, ecological crisis and world stock market collapse: global in scale and contingent events, devastating in their life-threatening intensity, infused within both social organization and the social modification of nature, and remote from the control of both individuals and states.

Environmental risks are in many ways the most insidious. The lurking threat of environmental danger is not only universal, but also undermines the confidences and certainties of modernity in the most mundane and telling ways. The unpredictability of the effects of technological interference in food chains, new forms of air pollution, shortages in basic water supplies and so forth, represent everyday risks to everyone's health and sense of security. Fears concerning climate change, ozone holes, extinction of species and other global environmental problems reproduce this anxiety on a more abstract level.

This more environmentalist approach to risk has been developed by Beck (1992a, 1992b, 1996), whose enquiry into the threatening nature of contemporary social life leads him to characterize postmodernity as 'risk society'. Beck's central thesis holds that technological advance is subverting the established logic and social acceptability of risks. Modern human-produced risks, he argues, were rendered acceptable on the principle of their calculability. They could be foreseen, to a large extent controlled and, decisively, insured against. Risks were limited in scope, time and space, and upon this assumption insurance mechanisms could be designed to assure the compensation of eventual victims. Postmodern risks, on the other hand, argues Beck, represent a radical break with all this. They are increasingly complex, unpredictable and uncontrollable within clear spatial and temporal boundaries. Public incredulity with scientific accounts of risk and the ability of the techno-scientific establishment to deal with them, has debilitated the bond of trust within society as a whole, and the effective protection afforded by insurance. In short, postmodernity represents a breakdown in both the technological rationality and social mechanisms of risk control.

Thus, when Beck speaks of the 'self-annihilating future', he strikes at the very heart of postmodern uncertainty, and rationalizes the environmentalist response to it. Not only does he attack instrumental reason and the power structures it engenders, but he also highlights the new social sensibility arising from the elevation of risk to global, unavoidable and unthinkable proportions. When Giddens (1994) turns his attention directly to the environment, he interprets it explicitly as 'modernity under a negative sign', the material expression of the practical limits of modernity. The existential dilemmas posed by radicalized modernity and manufactured risk, he argues, along with much environmental political thought, constitute the basis for political renewal.

Whilst intellectual critique deconstructs the philosophical and epistemological foundations of modernity, the environment appears to show modernity self-destructing in practice, with the result that the risk-strewn image of postmodernity is presented in a strikingly accessible and alarmingly proximate manner. As a result, the 1990s saw an explosion of interest in risk. The United Nations declared it the International Decade for Disaster Reduction; health scares over illnesses such as Aids, E-coli, 'mad-cow' disease, salmonella, and unfathomable new allergies and syndromes became a regular feature of the media and politics; and the journal *Living Marxism* vociferated growing public scepticism when it condemned the fabrication of an 'epidemic of fear' (Fitzpatrick, 1995). All of this sparked off a wave of sociological reflection on risk (for example, Royal Society, 1992; O'Riordan, 1995; Adams, 1995; Lash, 1996; Lofsted, 1997; Green, 1997; Franklin, 1998). Risk assessment and management had become both an industry and a responsibility incumbent on individuals in the conduct of their everyday lives.

What can be drawn from this contextualization of risk within postmodernity, that is, in conditions of risk as an integral part of everyday experience, is that the environment has become the medium of transmission of a general social malaise. In this context, it is not surprising that risk has been incorporated as a central dimension of institutionalized spatial practices. The formal environmental presentation of risk lies in natural disasters and the precautionary principle of sustainability. However, it is arguable that public acceptance of precaution and the general environmental turn in planning are based on a much more widespread perception of risk as an immediate, personal and quotidian challenge in everyday life. To this degree, risk easily inserts itself into or hovers over every dimension of spatial agendas: risks of energy consumption to climate change, pollution to health, development to biodiversity, transport systems to accidents, building technology to psychological well-being, settlement patterns to community, building arrangements to crime and safety, and so on. In such a postmodern socio-spatial environment, planning has become preoccupied with damage limitation; an arena in which the fears, more than the hopes, of society are played out.

Survival

The idea of risk society encapsulates the postmodern sensation of vulnerability. Out-of-control abstract systems contain and create unforeseeable dangers which are themselves far from abstract. In their environmental form, they impinge on the commonest of everyday activities: what to eat and drink, where to swim and whether to sunbathe, the choice of medical treatment, the use of household appliances, where to live, even breathing, have become hazardous activities requiring cautious decision-making. If that were not enough, contemporary environmental risks are disdainful of social and geographical divisions in the unfolding postmodern world order. Traditional strategies of insurance against risk, and risk avoidance through the protection afforded by class divisions and spatial isolation are no longer effective in the anarchic milieu of postmodern risks (Bauman, 1992b). This in itself would seem to have direct implications for urban planning, rendering obsolete one aspect of its traditional strategy in the area of public health, and thereby undermining an important source of legitimation. However, the implications of postmodern risk for planning practice need to be understood within a more complex picture of survival: the privatization of survival strategies and the new division of responsibilities between the individual and the state.

The roots of the privatization of survival have been identified by Bauman (1992b) as lying in modernity's belief in history and emancipation through which life became subservient to the grander Enlightenment project of the progress of humanity. Even human life, he argues, became a means to an end, a step on the path to a more perfect social order and the promise of greater future individual fulfilment. As a consequence, the religious sanctification of life gave way to instrumental reason, and transcendence of earthly existence through death began to assume its postmodern inversion: the transcendence of death through life. Immortality became a challenge of mundane existence, something to be strived for not through this-worldly existence, but in this world itself. The effect of this inversion, according to Bauman, is that:

> The horror of mortality has been sliced into thin rashers of fearful, yet curable (or potentially curable) afflictions; they can now fit neatly into every nook and cranny of life. Death does not come now at the end of life: it is there from the start, in a position of constant surveillance, never relaxing its vigil. Death is watching when we work, when we eat, when we love, when we rest. Through its many deputies, death presides over life. Fighting death is meaningless. But fighting the *causes* of dying turns into the meaning of life.
>
> (Bauman, 1992b: 7)

The modern conversion of death into a particularly private affair, and the fight against the causes of death into a life project, achieves a postmodern expression, argues Bauman, in terms of self-care. Attempts at transcendence of mortality might linger on through religious or human love, or totalizing common-cause ideologies (such as nationalism, socialism or racism), but the dominant, postmodern contribution is the privatization of the 'responsibility for the business of life'. The irony of the privatization of life, Bauman continues, is that it increases individual dependency, at the same time giving new shape to sociality. Whilst survival is affirmed as a private matter and private responsibility, the 'loneliness' it implies in a complex, risk-laden world radicalizes the cognitive requirements of expert knowledge (and therefore dependence on experts themselves), and the psychological need for sharing the lifestyle consequences of those decisions. Expert knowledge, however, is never sufficient, since its fallibility increases with its accumulation of specialist sophistication; it is continually changing course, opening up new doubts, discovering new dangers. In this respect Bauman takes up Giddens' theme of abstract systems and Beck's distrust of scientific knowledge and its institutions, but does not share their modernist optimism grounded in reflexivity. For Bauman, it is the task of experts to create a demand for their services by ever revealing new risks and showing the inability of individuals to deal with them alone.

At this point, the second feature of postmodern survival, concerning the redistribution of responsibility between the individual and the state, may be introduced. The privatization of the business of life was part of an ideological shift from the rights of individuals to have their health protected by the state, to the responsibility of individuals to protect themselves from risk. As risks became subjected to probability analysis, dissected into component factors and spatially mapped out, this expert knowledge is displayed for public consumption. Individuals are thus obligatorily equipped with the bare technical resources for assuming the role of their own risk assessment managers. Risk is not denied, but the prevention or minimization of risk becomes a matter of personal calculation and individual responsibility. With this privatization of risk, the implications of things going wrong are particularly bleak in postmodernity. As Green (1997: 119) explains, '[Modernist] rationality could provide no solace, or opportunities for revenge for such misfortunes, but it implied (at least ideally) no blame . . . [Now] victims of the failure of risk management may not be seen as malicious, but they are in a sense culpable, in their ignorance'. In other words, risk taking became a form of irresponsibility, negligence or wilful deviance worthy of blame and condemnation (Douglas, 1992).

In this way, a healthy lifestyle has acquired a moral dimension in the 'duty to be well' established by contemporary society (Greco, 1993), in which accidents and disease are perceived not so much as a misfortune, but rather as an indication of the moral quality of the individual. This moral dimension is, in turn, incorporated into concern for the health of the planet. The moral obligation to stay well physiologically becomes, simultaneously, a moral imperative to contribute to global ecological well-being, since the two are, in the postmodern preoccupation with health and survival, irremediably connected. However, the general integrity of an abstract proposition such as this is easily perverted through sustainability's technical agenda. Dominant physicalist discourse determines which issues are relevant, what kind of responses legitimate. As Grove-White (1992) observes, this has exclusionary effects for more existential perspectives on the environment, at the same time as alternative agendas are trivialized and ecological ignorance is denigrated.

From this perspective, the environment becomes an essential dimension of individual and ecological survival, the social space where individual responsibility, or irresponsibility, is brought into the public domain. Individual strategies of survival based on the elaboration of an appropriately healthy lifestyle designed to maximize the prevention or postponement of illness, encounter their limits in the socio-spatial environment. However much postmodern diversity allows for the selection of environments in carefully planned individual trajectories through life's socio-spatial *milieu,* the individual is eventually forced into confrontation with the deficiencies of other people's strategies which threaten to undermine his or her own.

The issue of smoking can be seen as a salient example of a trend which is spreading into all aspects of socio-spatial behaviour. The environment becomes the inescapable sphere of micro-politics in the sense of the medium of dispute over personal and global health concerns and, ultimately, the meeting point of postmodern self-interest and the vestiges of modern altruism.

The significance of all this for planning is considerable. Whereas planning could once decree, on behalf of the state, what constituted a healthy organization of the spatial environment, it can now only respond to a confusion of contradictory expert knowledge, conflicting interest groups and a plethora of lifestyle valuations. Given a situation in which the environment has become an important area for the public debate on health in a privatized personal and institutional landscape, planning is compelled to adopt the role of conflict mediation. From this perspective, the new role of planning as 'environmental manager', as proposed for example by Evans and Rydin (1997), depends not so much on the development of an appropriate knowledge and skills base of planning, or even the overcoming of the institutional strictures within which planning operates in order to be able to contain and perhaps progressively direct environmental conflict (Thomas, 1996). The real significance of the environment for planning would seem to be its mere existence as a problem area, upon which can be constructed the new conceptual and technical ground for the profession's legitimation as the institutionalized protection of the postmodern public health interest.

Ethics

The reintroduction of ethics and morality as contemporary issues is common to both postmodern and environmental discourse. Postmodernism does this through the re-examination of the social construction of human subjectivity; environmentalism through the reaffirmation of the individual as a natural entity in the larger world of life. In both cases, the individual encounters circumstances (both social and environmental) which demand the personal assumption of responsibility, and the adoption of some kind of justifying rationale for choice of lifestyle. Pluralism and enhanced individual autonomy imply that in postmodernity 'the agent is perforce not just an actor or decision-maker, but a moral subject' (Bauman, 1992a: 203).

Bauman develops this idea by arguing that postmodernity represents a movement towards the disintegration of mass society and the social adjudication of moral issues. Cultural diversity and recognition of the other, as part of the plurality of postmodernity, implies that 'agents' are forced to make moral choices as their own problem and responsibility. In the absence of unequivocal guidance (from church, state, community or family), 'individuals are thrown back on their own subjectivity as the only ultimate ethical authority'. Through this process of the privatization of morality, ethics has become 'a matter of individual discretion, risk-taking, chronic uncertainty and never-placated qualms'. For Bauman, the ethical paradox of the postmodern condition is that moral choice is accompanied by the deprivation of the comfort of universal guidance that modern self-confidence once promised.

This social condition is given a special urgency by the ecological crisis. The moral dilemmas of social behaviour and identity are compounded by the assignation of individual responsibility for the effects of lifestyle options on the ecological foundations of life itself. For Giddens (1994: 247), for example, the ecological crisis is the opportunity to 'remoralize our lives in the context of a positive acceptance of manufactured uncertainty', by forcing us to 'confront moral problems which were once hidden in the naturalness of nature and tradition'. As observed earlier, modernity separated man, the knowing subject, from the rest of nature, and laid it out at his disposal through its objectivization; scientific knowledge and instrumental reason made nature available as an object to be exploited, a resource. In this way, nature was surrendered to the general moral consequence of modernity, 'that gigantic exercise in abolishing individual responsibility other than that measured by the criteria of instrumental rationality and practical achievement' (Bauman, 1992a).

In this general respect, environmental thought mines a rich seam exposed by the postmodern erosion of the historical deposit of the Enlightenment. Hoyos (1989), for example, argues that the ecological crisis is not only a manifestation of the crisis of instrumental rationality, but also a profound illustration of the significance of the dialectics of reason; it is necessary to recoup 'the natural conditions of reason itself'. In Habermasian style, he proposes the need to articulate ecological responsibility with communicative action or, in other words, ecological consciousness with inter-subjective reason. In this way, the distortions of modernist subjectivity are doubly constrained, by both the social and natural origins of human reason. The 'world of life', as the context for the exercise of dialogical reason, would demand recognition and re-incorporation of the diversity and complexity of all (human and non-human) life in both political and ethical terms and, in so doing, recreate a sense of the 'poetic habitation of the earth'.

The postmodern critique of the self-defeating utilitarian attitude towards nature, made evident through the ecological crisis, inspires a moral interrogation of habitation, and thus of spatial organization and behaviour. In more concrete terms, it presents both a challenge and outlines a solution to contemporary institutionalized spatial practices. The challenge arises from the postmodern emphasis on diversity and difference, which problematizes planning's modernist foundations in mass society by undermining the legitimacy of global solutions to socio-spatial issues. Planning could no longer pretend to spatially organize society on the basis of universal values and lifestyles. In response to this dilemma, there was a resurgence of procedural theorizing which addresses the problem of diversity in terms of the need to establish adequate communicative practices (Healey, 1996; Forester, 1993); ethics became an issue for planning in terms of meaningful debate and understanding in a fragmented, multi-cultural and multi-racial social environment. This still left a serious substantive problem concerning the spatial project of planning, its material agenda. As questions of equality and redistribution were squeezed from planning's agenda, not only were planning's modernist spatial images made obsolescent, but also the ethical basis of planning's legitimacy as institutionalized practice. Its social value, what it was actually useful for, the specific contribution it could make through a superior spatial organization of society, were left in severe doubt.

In this context, planning's enthusiastic embrace of the environment can be seen as the opportunity to re-establish a moral dimension to institutionalized spatial practice. In the British context, a discussion of the dilemmas of planning practice allowed Underwood (1991), for example, to affirm: 'We all need a touchstone to maintain a sense of purpose in times of great uncertainty and stress. That is why I am proposing the maxim "the health of the planet is the highest good" – a third party interest we can all support'. Enthusiasm since then may have waned, but it still remains a proposition that most planners will not have renounced altogether. The environmentalist idea of stewardship of the earth as a moral responsibility was clearly an attractive proposition for reinvigorating planning's flagging moral fibre and social legitimacy. At the same time, the environment provided planning with the opportunity to modernize its social agenda. It had evidently found in the environment a new moral cause and ethical foundation whereby the defence of the public interest and common good could be plausibly posited in the conditions of postmodern spatiality. Urban environmental agendas cannot be fully understood outside this search for a postmodern ethical foundation of contemporary spatial management.

In other contexts, especially in the USA, ethical concerns arising from the environment had been forced upon planning by a more dynamic urban politics and the activity of grassroots organizations. The unequal distribution of environmental 'bads' such as toxic wastes and contaminating industries among poor sectors of the city provoked an environmental justice movement of considerable impact, in large part due to its connection to issues of race, ethnicity and gender (Harvey, 1996; Haughton, 1999). Here, the insertion of ethics into urban planning was through active politics rather than the more reflexive road taken by European planning supported by stronger state regulation of land use and environmental issues.

Sociality/solidarity

The material grounds for the moral and ethical dilemmas of postmodern culture have been associated with the post-war diversification of the social division of labour, the consequent erosion of class-based cultures, and the transfer of identity from the sphere of production to that of consumption (Heller, 1988). Moral and ethical issues, previously determined by class condition, have now to be resolved on the basis of personal choices concerning lifestyle, over which society imposes little authoritative guidance or sanction. The 'correct' or 'incorrect' conduct of personal relations, conformism or rebellion, 'good' or 'bad' taste and so on, lose their significance in soft postmodern culture. Individual choices have to be made in this context of radicalized relativism.

As has already been observed, the advantages of this ethical freedom of choice are offset by the 'loneliness' such individual moral responsibility implies. This is far from a merely abstract *denouement*, given the irrenounceable requirement of postmodernity to consciously construct personal identity, which in turn involves the taking of moral decisions with very practical consequences. The need to share such responsibility and its outcomes leads to postmodernity's 'lust for and invention of community' (Bauman, 1992a). According to Bauman, postmodern communities lack the assurance once derived from the legitimating claims to universal reason, and the solidness previously proportioned by the legislating authority of the state in a more stable social order. Bauman describes them as 'imagined communities', present as concepts rather than integrated social bodies, formed as a result of individual acts of self-identification and existing 'in no other form but the symbolically manifested commitment of their members'. Rootless in natural or material terms, communities are, as a consequence, irredeemably contingent and inherently precarious.

These forms of postmodern social affinity have been described by Maffesoli (1991) as 'neo-tribes'. Sensibility, sensation, sentiment and attraction are their supports, the 'ethical vectors for non-obliging association' in postmodernity. In emphasising the social character of aesthetic emotion and its function as social cement, he proposes his key idea of 'being together' as 'essentially a mystical reliance without any particular object'. Sociality as shared aesthetic experience is, he claims, the basis of neotribalism, the new form of solidarity in complex postmodern society. More recently, the invention of community has taken mediatized and virtual forms of the 'sharing of intimacy' through television reality shows and personal confessions, chat rooms and private cam-sites on the Internet, and so on, a general phenomenon in which any residual public interest is reduced to curiosity about the private lives of public figures (Bauman, 2002).

It would be an exaggeration to suggest that imagined communities, neotribes and virtual interaction have totally replaced social ties based on kinship, locality and class. What they do undoubtedly demonstrate is the fact that postmodern heterogeneity poses the problem of solidarity in a completely novel way. In the absence of claims to universal truths and amidst weak authoritative hierarchies, tolerance and respect for otherness become ethical concerns of vital political importance. However, they would appear insufficient in their very practicality. They only manage difference and its conflicts. The articulation of difference and the reconstruction of a sense of collective identity and common purpose become the more profound challenges facing heterogeneous postmodern society.

The political crisis of postmodernity, argues Lechner (1989), is precisely the inability of modern institutions, especially the nation-state, to continue to exert that integrating function they inherited from religion in traditional society. If religion once constituted the ultimate authority of a received order, politics was attributed this privileged place in the socially produced order of secularized modernity. It was given the responsibility of not only determining the conscious action of society on itself, but also the task of the representation of society as a collective order. However, the postmodern celebration of difference and plurality undermines the representational capacity of modern institutions. The modern nation-state in particular, with its monological, instrumental rationality and technological management style, loses both its appeal and significance with the increasing internationalization of economic, social and cultural life.

This political crisis arises not just from the fractioning of antagonistic class interests into a myriad of disconnected social movements and fragile communities. The key question for Lechner is whether modernity has lost its reforming impulse. The latter presupposes a belief in the possibility of greater future individual fulfilment and welfare through the intervention of society upon itself. However, on the one hand, the rapidity of contemporary change causes the future to dissolve into a mesmerizing, constantly changing present; on the other hand, the rational basis for reform appears increasingly inadequate, if not simply repressive.

The political attraction of the environment lies in the fact that it constitutes a universal reference for reconstructing a sense of community, solidarity and common interest in a socially fragile and fragmenting world. Ecology offers a paradigm for accommodating difference and articulating a new sense of totality. The ecological organization of life privileges variety, complementariness and local distinctiveness in a positive model of diversity. At the same time, it stresses interdependence, wholeness and respect as conditions of that diversity. In this environmental perspective, the imagined communities of postmodern sociality can be underwritten by the common biological interests of survival, and solidarity becomes an ecological exigency based on the biological rather than social dimension of human existence. The environment thus emerges as the area in which the fragmented sociality of postmodern society can be feasibly articulated, and where the separation of the individual and common good encounters the practical limits of its own contradiction.

In urban terms, two major consequences of this general political significance of the environment can be identified. On the one hand, the environment constitutes a representation of space in which the common interest can be reasonably posited, and therefore the environment can be credibly presented as the basis for the construction of a sense of solidarity amongst citizens. Upon this appeal, institutionalized spatial practice can resurrect a claim to operate in the public interest as a vital factor of legitimacy. On the other hand, it also transforms the type of urban project which can expect the support of urban society as a whole, and capture the public imagination in such a way as to acquire effective political viability. The spaces of networks and flows which interconnect and articulate diversity and difference become increasingly important. This prioritization of flows and channels is central to the explorations of Castells (1996) in relation to information and communications technology. However, important as this area may be to urban development and the production of new spatial patterns, it is concerned with the production of virtual spaces of an as yet uncertain political significance.

In contrast, the environmental focus on the same theme is much more embedded in the general social experience of space. Castells (1997) himself has argued that the environment is a key factor in the struggle for control over space, in the opposition emerging between two spatial logics: the virtual space of electronic flows and the real space of places in which people live out their everyday lives.

The undeniable fact seems to be that, in contemporary society, the environment irrevocably mediates social contact – real or symbolic, and frequently conflictive – between diverse urban groups and cultures. In terms of professional practice, the environmental version of flows and channels as spaces as collective interest has the effect of contributing to a resuscitation of urban design, in its attention to the quality of public space. What was a residual concern of modernist planning, with its emphasis on buildings and uses, becomes a privileged area of postmodern planning. In this way, urban environmentalism renovates interest in streets, squares and open space networks, now politically reinvigorated as spaces of the shared consumption of natural resources, as well as prioritizing the spaces of flows proper to natural resources themselves, such as rivers and streams, woods and waterfronts and natural landmarks. In short, it can be argued that urban design has become a leading edge of contemporary professional practice, in part at least, in response to the need to spatially represent postmodern social solidarity based on the environment.

Individualization

A recurrent theme in the previous sections has been the increase in demands put on individuals. Whether it be giving sense to one's place in the world, assessing risks and taking (or not taking) action, making decisions of an ethical kind or coming to terms with how to establish and manage social relations, the individual is obliged to make decisions and assume personal responsibility for the outcomes and implications of those decisions. In the conditions of 'reflexive modernization' (Beck *et al.*, 1994) we have tried to show how the environment is an important frame of reference for individual decision-making and how urban planning theory and practice has assimilated this dimension of socio-spatial existence. We shall now look at individualization in more detail and in the light of processes of social change which make individualization and personal choice a social obligation.

Recent sociological inquiry has given individualization a particular meaning. It is not concerned with Enlightenment ideas of self-realization, of the individual becoming him or herself through knowledge of the world and reflection on his or her role within it; nor is it to be confused with the capitalist version of the enterprising 'self-made man'. Individualization has more to do with the construction of identity through the processes by which people are integrated into the social order (Bauman, 2001). Until recently, individual identity was determined by relatively stable structures such as class, kinship, religion and tradition, which in turn outlined a person's position within the field of work and the opportunities for the development of skills and a career. In this 'work society' with its 'work ethic', identity arose out of social position and was built on through personal effort, consistency, loyalty and virtue. Experience over the past two decades or so has demonstrated conclusively the obsolescence of these qualities during economic restructuring and the emergence of 'consumer society'. It is not just that religion, tradition, class and family – and the state structures which supported them – have all been weakened, but that work has been both radically restructured in itself and displaced as the primary source of social identity (Bauman, 1998). Skills and jobs are no longer for life in the volatile, flexible labour markets of the global economy involving continual innovations in technology, company organization and product development. Amid this indeterminacy and instability, work is now located at the intersection of flows rather than solidly fixed in structures (Lash, 2002) and the opportunity for self-reflection disappears for the contemporary individual in a permanent state of agitation, perforce connected through the internet, email, the mobile phone and cable television in order to network alliances, coordinate social commitments and keep abreast of at least some of the proliferating political and cultural events.

In consumer society, work is no longer sufficiently important in production terms or stable enough in a personal sense to provide social identity. 'What do you do?', when referring to occupation, is no longer the most significant question when inquiring about others. Intensified consumption is now what provides identity, integrates people into society and moves the economy, though as Bauman (1998: 30) has pointed out, in a new way: 'Producing' is still a collective endeavour and shared experienced; even in flexible working conditions it presumes the coordinated division of tasks, the cooperation of actors and some face-to-face communication. For its part, 'consuming' is a 'thoroughly individual, solitary, and in the end lonely activity; an activity which is fulfilled by quenching and arousing, assuaging and whipping up a desire which is always a private, and not easily communicable sensation'. The need to communicate aesthetic interests rather than share work-based ethical norms is the driving force behind 'imagined communities' and virtual contact, and neoliberalism is its official policy.

When Margaret Thatcher proclaimed that there was no such thing as society, she was partially and unwittingly right, in the sense that modern social institutions were to become insufficient to provide the resources for individual identity and life-support. Of course she was mainly interested in dismantling certain state apparatuses, organized labour and the political interests behind them, and in promoting an entrepreneurial individual more akin to the idealized self-reliant figure of nineteenth-century capitalism. Other facets of individualization such as moral relativism, declining national pride and patriotism, the subversion of the family and so forth, alarmed her. In Beck's influential work on individualization, he is at pains to stress that 'institutionalized individualization' is the very opposite to the ideological notion of the self-sufficient individual. Individualization, he insists, has historically been and is today a form of socialization, and that the individualization processes of 'second modernity' are a demonstration of self-*in*sufficiency, in that 'human mutuality and community rest no longer on solidly established traditions, but, rather, on a paradoxical collectivity of reciprocal individualization' (Beck and Beck-Gernshein, 2002: xxi).

The implications of 'institutionalized individuality' for the city are considerable, for in consumer society it is not just individual identity but also city images that need to be reconstructed accordingly. The old urban images of work-based industrial cities become as redundant as the workforce it once employed. To be a competitive city within consumer culture means responding to and amplifying the offer of material goods, cultural events and spatial opportunities for the aesthetic exchange between individualized citizens. It is not just a question of satisfying consumer demand and the desires of citizens, but the obligation to materialize this new sociality in the very image of the city. Maximum consumption rather than full employment becomes the main policy goal and legitimating factor of city governments, since individual identity and citizen identification with their city depend on it. The low-paid and out-of-work, or the 'flawed consumers' of contemporary society are then relegated to the status of second-class citizens in the sense that they fail to live up to the requirements of the economic structure and social demands of the consumer city. Indeed, their own classic cultural forms, such as the local pub or football team, may either wither away or be expropriated by the expansive demands of insatiable, fully-fledged urban consumers. The restructuring of urban economies around retailing and the culture industry is exclusionary not so much as a question of class refinement but more in the sense of the restructuring of the very terms of citizenship: disposable income and the desire to consume become the principal considerations. This puts the earning and consumer habits of the debilitated working class, the old and the poor in a systemically disadvantaged position. In the spatial and temporal segmentation of urban life, it is the enthusiastic young consumer and professional groups with both economic and cultural capital who are the most willing, able, enthusiastic and demanding of opportunities to construct their identity through the consumption of new urban space, and at whom this space is most obviously aimed.

It is no surprise that the policy coalitions behind urban restructuring are strongly influenced by the service sector business community and surveillance organizations of the consumption economy and its policing, which oversee city centre revitalization and the development of cultural quarters, entertainment zones and 24-hour activity and so forth, within careful limits and strict controls. In this context, the environment is assimilated into the logic of consumption. The built environment, especially its historic components and natural elements and features, are integrated into the city as a product to be marketed. Urban ecology is reduced to the general requirement of cleanliness and security in the moment of individual consumption. However, there is always the danger in urban space that the flawed consumer – the poor and marginalized – will disturb these settings and spoil the act of consumption though their mere presence or the enacting of some kind of symbolic protest or act of violence against their exclusion. In the final section we examine the politics of the city as a space for individualized consumption.

Urban integration and governability

Individualization is a generalized but uneven tendency within contemporary society and tends to disconnect personal from collective well-being in the midst of processes which accentuate material inequality. Individualization brings people together precariously in the act of consumption (in the shopping centre, concert hall, festival, club and bar) in such a way that profound social differences are simultaneously confirmed and momentarily erased in this particular form of socialization. It would therefore seem to be of great political significance for cities and their management in the sense that individualized consumption – and not employment, basic services and welfare provision – constitutes the primary responsibility of city government and the 'cement' of urban social life, somehow managing to contain and suppress the tension, frustration and potential for conflict bubbling underneath the urban surface.

There are at least two useful ways of understanding this new dimension of urban governability and its connections with the environment. First, it can be approached from the perspective of the emotional constraints and socializing limitations of individualization. The neoliberal ideology of self-sufficiency, given credibility through consumption, impetus through the reform of the welfare state and obligation through the labour market, is based on the exercise of supposedly unrestricted freedom of unlimited choice. However, as we have seen, the incessant and obligatory exercise of choice reveals what Beck has called 'the fundamental incompleteness of the self'. However, he argues that individualization needs to be accepted and understood as the core of individual and political freedom in 'the second modernity', in the sense that the isolation of individuals within their own social groups provides the basis for new forms and direction for political association and action. In particular, he casts doubt on the ability of nation-states and governments to be able to deal with the complexity of globalization any more than individuals can, and suggests that it is at the local level of towns and cities that sense can be made of its bewildering processes and contradictory responsibilities. In a new cosmopolitan local democracy constituted by individualized individuals, the conflicting demands of different policy areas and the direct experience of them could lead to 'a repoliticization of municipal policy, indeed a rediscovery and redefinition of it by mobilizing programmes, ideas and people to make the incomprehensible and impossible real and possible, step by step' (Beck and Beck-Gernsheim, 2002: 168).

A second and less optimistic account of individualized inequality revolves around the question of the disappearance of any sense of mutual obligation. If individualization means that one's life has to be individually constructed, a personal 'biography' written and enacted in a way which is substantially independent of other people, then this obligatory self-centredness leads inevitably to a certain indifference to the fate of others. More than that, there is always the danger of a latent resentment towards others, degenerating into contempt for those who fail, for one reason or another, to even attempt to play the demanding rules of the game. The latter are regarded as deficient, unworthy of compassion and the helping hand of society, deserving at best only help to help themselves onto the road of tremendous effort that normal citizens have to make. The effort and stress of survival in consumer society, and its never-ending material and psychological insecurity, is returned with venom on those who, for whatever reason, remain outside its embrace.

Bauman's (1998) analysis of the rise of the underclass is telling in this respect. The underclass (a mixed bag of the homeless, single mothers on welfare, petty criminals, beggars, long-term unemployed, those 'refusing' work, school drop-outs, alcoholics and drug dependents, illegal immigrants, teenage gang members, the list changes and expands) came to be defined as those not at the bottom of society (the 'working class') but those *outside* society, fulfilling no useful function and essentially dangerous. Convulsed by its own anxieties, continues Bauman, normal society came to fear and stigmatize the underclass. The remedy, along with zero-tolerance policing and criminalization, was the work ethic now discarded by consumer society as the principal form of social identity and integration. However, more than a stepping stone into the consumer normality, argues Bauman, the success of welfare-to-work programmes should be measured not so much on how many people they get into work, as the disappearance of any sense of mutual obligation or moral engagement with their plight:

> Let me repeat: in the beginning, the work ethic was a highly effective means of filling up factories hungry for more labour. With labour turning fast into an obstacle to higher productivity, the work ethic still has a role to play, but this time as an effective means to wash clean all the hands and consciences inside the accepted boundaries of society and the guilt of abandoning a large number of fellow citizens to permanent redundancy. Purity of hands and consciences is reached by the twin measures of the moral condemnation of the poor and the moral absolution of the rest.
>
> (Bauman, 1998: 72)

However, absolute moral disengagement with others is not possible in an urban setting. However much the streets are patrolled and equipped with surveillance systems, however much residential security is achieved and behavioural patterns adjusted to 'safe' times and places, the chance of coming face-to-face with the poor and non-integrated is always imminent. Nor is moral disengagement an acceptable political proposition. In consumer society, widening the opportunities for consumption is the principal legitimating factor of city government but it contains its own moral void which inclusionary urban policy fails to adequately conceal through such things as partnerships and regeneration programmes. Neither can novel techniques of public administration such as opinion polls and focus groups, publicity campaigns, information availability, performance monitoring and online access to public services.

We argue that the environment fills this moral gap in the politics of urban management, but in a symbolic sense. It demonstrates that indifference is not absolute or manipulative, without having to encourage direct contact between different social groups or promote interpersonal commitment. It sets social relations in terms of a global community, allows the city authorities to address the population of individuals as a species 'we', landscapes the biological 'us' and extends civic responsibility to those excluded from full memberships of urban society. In this moral legitimation strategy lies, surely, the political appeal of the environment to urban managers.

Conclusions

This chapter has tried to draw out the more salient connections of environmentalism to the cultural condition of postmodernity, and demonstrate the relevance of the issues raised to the understanding of urban environmental agendas. The underlying argument has been that the inclusionary, totalizing dimension of the environment derives not so much from the conceptual structure of environmental thought, as from the experiential conditions of contemporary socio-spatial existence. It is this which authorizes the general social significance of the environment, and in so doing makes it an attractive alternative to the destitute modernist foundations of institutionalized planning. The environment is not, of course, the only important issue in contemporary planning practice, but it does tend to encompass all the others, to such an extent that it is reasonable to suggest that the environment has become a new paradigm for the management of space, in that urban environmental agendas appear to contain at least the seeds of a formal, universal model of urban practice. However, what emerges is not so much a new planning dogma (and its concomitant master plan) but, rather, a new set of foundational concerns which outline a change of direction for planning, a new route to be negotiated through the complex map of a fragmenting social order. The main propositions can be synthesized as follows.

First, and at a general level, it may be concluded that both environmentalism and postmodernism deconstruct modernity, and express the sensibility of postmodernity in strikingly similar and complementary ways. Where they most obviously differ is in their conception of historical discontinuity. Postmodernism compresses and reconstructs time in the present; environmentalism expands time and reconstructs it within ahistorical, evolutionary horizons. Postmodernism abandons any practical sense of the future and of a collective project; environmentalism places the future, the very possibility of a future and the collective responsibility for that future, to the forefront. Does this imply that environmentalism, within the general sensibility of postmodernity, constitutes a radical paradigm shift, and if so of what kind? This is certainly what environmentalists call for in relation to the issue of development, and it is tempting to suggest that this is exactly what is happening in wider cultural terms. The argument would run that the transcendental foundation of traditional society was succeeded by the instrumental rationality of modernity, in turn to be replaced by the ecological reconstitution of the currently tattered, strife-torn and degraded postmodern global village: the disintegrating social order of postmodernity finds its road to salvation and renewal in the planetary embrace of the natural order of the universe. This is a question which must necessarily be left open.

Second, environmentalism is firmly embedded in the postmodern social experience of space and time. Risk and insecurity are integral dimensions of contemporary cultural experience, yet it is in the environment that risk and insecurity acquire a sensuous, material and, above all, potentially manageable form. Whereas the health of the planet has become an abstract global issue addressed by the environmental sciences, the general interest in the health of the environment can be seen to stem from purely social processes which oblige individuals to interpret their personal health and well-being in terms of the socio-spatial environments they inhabit in their everyday lives. Whilst the environmental problem appears to demand a reconsideration of ethics and the question of solidarity between individuals, groups and nations, the willingness to accept the general premises of sustainable development seems to grow out of the practical dilemmas arising from specifically postmodern forms of sociality. All this is not to suggest that biophysical environmental problems do not exist, and much less deny the seriousness of certain effects and consequences. But it does establish a complex set of relations which mediates the ways in which such problems, effects and consequences achieve social importance and acquire certain forms.

Third, in these very concrete ways, cultural experience inclines towards the formation of a spatial consciousness defined in terms of the environment. However, it does not imply that this consciousness is rationally and coherently developed on the knowledge base of ecology, as dominant environmental discourse likes to insist. In fact, ecology would seem to be irrelevant to the formation of environmental consciousness, except in its most superficial, mass-mediated form of seductive images of the remains of wildlife and dramatic predictions of 'natural' disasters. On the contrary, environmental consciousness seems to arise from multiple facets of social experience which connect in a complex and irregular way with the formal components of urban environmental agendas. It is therefore hardly surprising that the environment, more than a single coherent issue engendering consensus and decisive political action, is more an area of conflict arising from the epistemological, ontological and experiential diversity of ways of understanding it. However, seen from a cultural perspective, it is above all an area of spatial dispute from which there is no obvious escape.

The idea of a cultural politics of the environment has formed the basis of some optimistic speculation on the future but, as seen in Chapter 1, this tends to undertheorize the structural dynamics of globalization, for example the neoliberal ideology of contemporary capitalism and its distorting effects on local/global cultural intersections. Similar observations can be made with regard to postmodernity. Postmodernity, like modernity before it, is not a stable and unequivocal condition, but rather a way of describing certain dominant features in the movement of society. Now, into the twenty-first century, both postmodernity and environmentalism have become integrated into social practices and routinized in daily life. As a consequence, the general features of postmodernity described in this chapter have undergone subtle shifts of emphasis. In the case of risk, for example, the social profile of some areas of risk may have diminished (such as those surrounding employment and sexuality) whilst others have sharpened (fear of crime, the untrustworthiness of financial institutions, terrorism). Similarly, risk management strategies have been both developed (for example more extensive networking, permanent interpersonal electronic communication, the elective reconstruction of the family, the intensified control of space), and in other cases abandoned, for example as more and more people fail to take up personal financial provision for old age, assume high debt and bankruptcy as a positive personal financial strategy or embrace danger through adventure sports. Survival appears as an increasingly individualized affair, simultaneously rational and haphazard, whilst solidarity and the propensity for political engagement show signs of a revival outside institutionalized forms, for example the anti-war demonstrations over Iraq and a greater disposition, in the appropriate circumstances, to pay taxes, whilst charitable donations decline and the fall into prejudice is always just round the corner. The social landscape of postmodernity is in perpetual change.

If postmodernity is unstable, so too is it unevenly distributed. It presumes skills and resources which frequently exceed the motivational sources and economic capacity which, in particular, the old and the poor need to be able to integrate into society; in developing countries postmodernity describes the conditions of the local elites much more than the majorities surviving in the informal sector; and postmodernity takes little account of the persistence of cultural traditions and premodern social practices. Nevertheless, it is omnipresent in some form or another, irrigated by global institutions, economic organization, tourism and the media to such an extent that material survival, cultural understandings and social lives are affected one way or another by the space-time compression of postmodernity.

Chapter 4
Discourse, power and the environment as urban ideology

Introduction

The preceding chapters developed three major areas of concern: the relation between sustainable development and the globalization of capital, the spatialization of this relation through urban environmental agendas, and the sensibility of citizens to environmental issues through the social and cultural experience of space. We intimated the relations between these three fields through key analytic concepts such as the ecological modernization, neoliberal urbanization and social regulation, whilst at the same time indicating the logical inconsistencies and practical limitations of urban environmentalism in both social and ecological terms. Whilst lack of results is frequently the cause of frustration and dismay among planners and environmentalists, our interest has focused on what might be called an evident 'success' of urban environmentalism: the political currency of ideas on the environment, especially in transforming the propositions within which space is debated, appropriated and administered. The aim of this chapter is to provide some theoretical order to these themes.

The sense of success of urban environmentalism that we are insinuating is expressed in part by the fact that the environment has become fully integrated into the policy agenda as a question worthy of the attention of government and requiring administration. The institutionalization and technification of environmental issues reflects the waning of environmental politics, with the environment no longer inspiring the critical thought, international agreements, innovative legislation, alternative technological and cultural projects, political movements or public attention it engendered towards the end of the last century. In urban terms, the search for sustainable cities appears to be yielding only minor improvements to their 'unsustainable' predecessors. Urban environmentalism has become mundane and routine, turned into a field of obligation rather than liberation, of moral conflict and practical dilemmas, occasionally enlivened by spectacular celebrations of nature in the city but on the whole a difficult and dispiriting challenge amongst so many other pressing matters in the fight for survival, whether global and economic or cultural and individual.

The sense of success is also implicit in its achievement through ecological modernization and the prevalence of the interests of global capitalism. We have argued that nature and the environment are not atemporal, external entities but contested social constructions in both a symbolic and material sense. Environments are therefore historical, unable to be captured in their full significance outside the general movement of society and the organization of integrative systems of symbolic and material production. Here it is useful to return to Acselrad's (1999) reminder that sustainable development is not an analytic concept but, rather, concerned with social practices aimed at the future but enacted in the present. The importance of establishing the authoritative voice when talking about and in the name of the environment suggests the need to address more explicitly the question of how this authority is established through discourse on the environment, as well as the issue of power and its exercise in urban space. In this chapter we therefore begin by exploring the notion of discourse more fully and how it can be related to social change in general and urban environmentalism in particular.

Discourse and the shaping of urban reality

Our experience tends to confirm a limited interest in and often hostile reaction to the notion of discourse on the part of professional practitioners. This is understandable from various points of view. Professional experts are concerned with concrete problems which they are obliged to resolve rather than problematize, usually within pressing schedules. Their institutionally structured fields of competence, along with the competitive pressures to respond to the demands of public employers or private clients, combine to make self-reflection on what they are actually saying through their discursive production (technical reports, policy statements, advisory documents, development proposals, administrative decisions, and so forth) a potentially uncomfortable or unprofitable activity, aggravated by the level of abstraction of much academic work on discourse which can be off-putting and seen as not worth the intellectual effort.

Nevertheless, academic interest in the fields of planning, urbanism and environmental studies has grown enormously over recent years. This is often justified by the claim that academic research can benefit the professions themselves by providing a sharpened awareness of the discursive nature of their respective fields of knowledge, leading in turn to a more critical understanding of issues, better communication between professions over complex issues and more socially progressive professional practice. Attention to discourse, it is argued, can increase our understanding of 'an arena of constant struggle over meanings and values in society, played out in the everyday micro-level practices of planning' (Richardson, 2002) and 'expand our practical tools, sharpen our critical judgement and widen the circle of democratic discourse' (Sandercock, 2003), since discourses are held to frame the possibilities of thought, communication and action for practitioners and theorists alike.

We have considerable reservations in this sense, but subscribe wholeheartedly to the proposition that critical attention to discourse provides valuable analytic insights and can improve the understanding of social practices in general. Discourse analysis involves taking a careful and systematic look at how issues come to be defined, constructed in particular ways and given prominence, how other issues and understandings of events are 'talked out', the interests and power relations in play, and so forth. The critically-minded professional will often have an intuitive appreciation of these sorts of process and, besides, discourse is not the only way of exercising or understanding power, although the question of 'non-discursive' practices will be left until later.

The importance of discourse

Interest in discourse derives from the general proposition that language plays an important part in the constitution of reality. In a similar way to the constructivist approach outlined in Chapter 1, it is held that reality is not an objective external entity which language merely represents, but rather that language mediates actively in providing form and giving meaning to reality as it is actually experienced. The precise nature of this relation has been a central theme of contention in linguistics. At the extreme, it has been argued that if reality cannot be conceived without some kind of linguistic intervention, is reality nothing more, in social terms at least, than a product of language or reducible to mere language itself? Against this radical constructivism, a more accepted position nowadays is that language mediates in the production of multiple realities, so that Macnaghten and Urry (1998) could declare the end of the 'sterile debate' over constructivism and realism and enable attention to be directed towards the social practices within which language is employed.

From a complementary viewpoint, interest in language can be seen in relation to postmodern change and uncertainty, in which reality would seem to have no fixed coordinates, no firm underpinnings, no reliable systems of stability or teleology. Recourse to the relatively stable constructs of language implies something to cling on to, especially for those for whom language is their main tool, such as academics and, increasingly, planners. On the other hand, language clearly does have a relation to social and cultural change, and it has been argued that the interest in language reflects its increasing importance in bringing about such change (Fairclough, 1992). Whilst acknowledging that since the nineteenth century Marxist social theory in particular has highlighted the importance of language, Fairclough argues that there has been a significant shift more recently in the social functioning of language, 'a shift reflected in the salience of language in the major social changes which have taken place over the last few decades'.

By way of illustration, anyone who experienced the transformation from Fordist welfarism to neoliberalism can easily appreciate the importance of language. Most evident in its radical form in the UK and USA in the 1980s, social change involved the deployment of a strident and aggressive political language to demolish old assumptions, impose new 'common sense' attitudes and expectations and so discursively reshape the contours of the future. This did not involve sophisticated argument but the blunt 'get on yer bike' type of discursive tactics of a Norman Tebbitt (a minister in Margaret Thatcher's first government telling people what they should do in the case of joblessness). It was this sort of political discourse imposing the values of private enterprise and the market which prefigured the technical discourse now dominating urban planning and the physical change of cities, with concepts such as competitiveness, attractiveness, quality and sustainability now often appearing, misleadingly, as a natural evolution in urban thinking.

The expansion and diversification of mass communications have also been important in this respect. Discourses are now disseminated instantaneously to global audiences with a combined audio-visual effect through the mass media, advertising, the Internet and the latest generation mobile phones. These new technological systems clearly present enhanced opportunities for promoting social change through talking about, and therefore establishing particular understandings of, phenomena such as development, security, social and individual well-being, cities or the environment. Indeed, sustainable development is a good example; it does not actually exist and has no precise meaning, yet through extensive media coverage of environmental ills and striking images of a ravaged earth, most people now have at least a faint grasp of some new set of meanings circulating around the word 'sustainability'. The work of scientists and activists could hardly have had such widespread effect without the modern media, as organizations such as Greenpeace are only too aware.

In contrast to linguistics, discourse can be understood as both words and images, with not only text but the full set of communication and information technologies playing an important role. In the case of risk, for example, Beck (1996) insists that risk society is always a knowledge, media and information society at the same time. With respect to television news coverage Cottle (2000: 43) found that television programmes trade both commercially and symbolically on cultural sensibility towards nature, registering and giving voice to the explosion of environmental awareness but providing 'precious few opportunities to elaborate a form of "social rationality", much less directly engage the discourses of institutional authorities charged with the management of environmental risks'. The media, by emphasizing the spectacular and aestheticizing the environment, produce an informed but unknowledgeable audience receptive to the technical presentation of environmental issues but ill-equipped to deal with them. In contrast, local cultural understandings of nature and the environment constitute knowledgeable but technically inadequate appreciations of environmental problems.

Discourse can be seen as a form of organization of this plurality of knowledge bases and social interests through concrete social practices, with specialist institutions planning playing an important role. The general point to be made is that in times of systemic change such as neoliberal transformation, the role and importance of discourse becomes more visible. New understandings and the re-elaboration of old beliefs have to be constructed through language in order to facilitate and legitimate new social dynamics.

Discourse in planning theory

Discourse analysis is a major dimension of social enquiry with a strong influence in both planning and environmental research. It emerged in the post-structuralist mêlée of the social sciences in the 1970s but did not have much effect on urban studies and planning theory until the 1990s. Interest in discourse is generally considered to be part of the 'reflexivity' of postmodernity, or the urge to re-examine the assumptions, knowledge bases and social practices of the applied sciences and professional disciplines. For planning, this meant developing a critical look not at the economic and social structures within which planning operated (as developed by Marxist critique), but a more introspective questioning of planning's own field of expertise, the type of knowledge it privileged, how this was deployed and with what effects.

As planning practice clutched eagerly at the environment for some revived substantive agenda and progressive social function, planning theory embraced discourse analysis with equal enthusiasm. Evidently, the notion of discourse provided a convenient new focus for planning's reflection upon itself – its social function and operational characteristics – by providing some much needed intellectual weight to counterbalance its debilitated institutional status. Neoliberal transformation had forced planning into an 'enabling' and 'facilitating' role, although what it was enabling and facilitating often led to depressing physical and social landscapes. By highlighting the discursive nature and practices of planning, discourse theory provided a more removed and sophisticated version of the processes underway, allowing planning to see itself as the organization of meaning and communication rather than concrete change. As such, discourse theory allowed planning to both simultaneously relieve itself of responsibility and seek relegitimation.

Modernist planning had been firmly entrenched in technical knowledge provided by positivist social science and its application through state institutions. Discourse theory provided the means for a critique of the legitimacy claims of planning based on this technical rationality (Rydin, 2003; Fischer, 2000) through a new approach to the old argument that in the real world there is no such thing as a purely technical decision and that rational argument is always undertaken with respect to particular interests, acting as a cloak for the operation of power (Flyvbjerg, (1998). Since interests are unevenly distributed throughout society, then *the* rational argument ceases to exist as something objective, universal and attributable to the state; in its place there has to be recognized a multiplicity of arguments with diverse rational foundations and social interests. In this context, the 'rational argument' is reinterpreted as a 'discourse' with only 'claims to truth'.

This provoked divergent theorizing of planning, frequently characterized in terms of Habermasian and Foucauldian schools. On the one hand, with interest focused on rationality, or the multiple rationalities of real world situations, the problem for planning was defined as one of communication, in the sense of the need to improve the understanding of complex phenomena by creating spaces for interlocution and reasoned debate. Habermas' theory of communicative action provided the inspiration for communicative or deliberative planning theory, a normative approach aimed at achieving consensus and the 'best solution' through rational dialogue (Forester, 1993; Healey, 1996). The assumption behind such theory is that consensus is possible in adequately defined and organized situations, similar to the 'ideal speech situations' formulated by Habermas. On the other hand, with the focus of interest on power, situations of this sort were rejected as idealistic and inapplicable to the sorts of issues dealt with by planning. Particular interests, it was argued, would always distort not only rational argument but also the conditions under which competing rationalities engaged in dialogue. In Foucauldian fashion, it was held that power is everywhere, including in rationality; this was the 'dark side' of planning which communicative idealists failed to acknowledge (Flyvbjerg, 1998).

Discourse analysis, planning and social change

The kinds of theorizing outlined above provide normative and critical accounts of the planning process whilst paying less attention to substantive issues. In general, planning's new spatial agenda of the environment and sustainable development is often taken as a given or some kind of fortuitous apparition. Whilst planning theory reverts inwards on itself, we argue that it is necessary to examine the relations between this type of theoretical and professional practice and the conditions under which they emerged. This means going beyond the previous observations concerning the substantive redundancy of planning and its search for legitimacy. For our purposes it requires, above all, seeing planning as a social practice concerned with the administration of space, and the environment as a particular conception of space.

Using discourse to understand social change always involves, schematically, some form of triangular analysis of language, the context in which it is used and the social effects it produces. Whilst what and how things are said and written, the grammatical devices employed and so forth, are of inherent concern, that which distinguishes discourse from other types of linguistic analysis is the central place assigned to the circumstances and effects of the use of language: discourse is understood as a social practice or 'language in action'. Discourse may therefore be seen as concerned with the construction and mobilization of meaning in specific historical contexts and social situations. However, discourse theory and analysis embrace many different disciplines and as tools of enquiry contain many different methodological options (Sharp and Richardson, 2001). A thorough account of discourse is beyond the scope of this book and the following sections simply outline some of the major propositions and avenues opened by discourse analysis, with a particular interest in exploring how environmental knowledge has been constituted, incorporated into urban planning and with what effects. As with much critical study of discourse and the spatial issues of planning, the work of Michel Foucault will be used as a general reference and guide.

The linguistic emphasis

We begin with some brief observations on the linguistic side of the triangle which concerned Foucault very little but which has been a major concern of environmentalists. Linguistic approaches to discourse analysis emphasize the dynamic social processes through which the meaning of words is contested, and how the lexical and grammatical resources of language are employed to structure communication and social interaction in specific contexts (Fairclough, 1992). On the one hand, linguistic analysis draws attention to how the grammatical devices of language are used to naturalize events, avert or assign agency, create positions of authority, and so forth, and can usefully be applied to the analysis of communicative practices in expert meetings and public participation exercises. On the other hand, linguistic analysis focuses on the inadequacy of words to describe something in sufficient detail, accuracy or truthfulness. The word 'sustainable', for example, is now used in multiple ways, often simply to indicate the economic viability or physical durability of something over time, playing on but distorting its ecological significance in sustainable development discourse. Similarly products, including cities, may be green-labelled as a marketing strategy.

The imprecision of language is the basis of a call by Harré *et al.* (1999) for a new lexicon in order to promote effective environmental change, in a fashion comparable to the terminology devised by Marx in order to describe a new way of understanding capitalism. In their critique of existing language Harré *et al.* argue that orthodox linguistics, which sees languages as abstract autonomous systems of signs, is inadequate to capture meanings that are always open-ended and embedded in the concrete and complex activities of everyday life. Words, they claim, can have no definitive semantic definition or exclusive rules of use; rather, words and their use always depend on who uses them and for what purpose in any particular context.

Their 'integrationist' approach places importance on the interpretation of what is being spoken or written, and the 'accreditation' of the discourse. Accreditation refers to the mutual acceptance of meaning by diverse participants in any particular circumstance, and is a particularly useful notion when participants from different discursive communities (science, the media, government, business, the professions, and so forth) are involved, as is almost always the case when talking about 'cities' and 'environment'. Accreditation is vital in establishing 'truth effects', and scientific knowledge with its universalist pretensions plays a key role in this sense by giving weight to general arguments by incorporating the nominal use of scientific theory and measurement. From a linguistic angle, Harré *et al.* (op. cit.) point out that the authoritative status of scientific knowledge is used not only to describe the world as it is but also to establish how the world ought to be. This is achieved through rhetorical persuasion and the use of metaphors both within scientific discourse and by using science as a metaphor in itself when employed in a decontextualized and ungrounded fashion.

They argue, furthermore, that the metaphor becomes a vital communication device. From their perspective, metaphor is more than a displacement of concepts; it integrates two conflicting accreditation values. By saying that one thing is 'as if' it were something else, it bridges the gap between discursive domains and creates at least a provisional understanding between, say, different branches of science, scientist and layman, urban planner and environmental engineer, city mayor and city investor. Their concern is not, therefore, with the metaphor as figurative meaning, but with its communicative function.

This communicative function of the metaphor deserves further comment. Harré *et al.* (op. cit.) explain that orthodox linguistics treats words as surrogates for facts or sets of facts, a parallel presentation of the real word through signs. From this point of view, words represent the world as it is. This is, generally speaking, the scientific mode of discourse. By contrast, nonsurrogational discourse is moral and aesthetic, and addresses the world as it ought to be. There is thus an evaluative dimension to nonsurrogational discourse. In this sense, metaphors are an especially neat linguistic manoeuvre, since 'they lead us from an "as is" to an "as ought" by a route that does not seem to depart from the surrogational mode'. Take, for example, the term 'sustainable city'. It is based on a loose assembly of arguments and data (on energy flows, production and consumption patterns, waste and pollution, lifestyles, and so on) which purport to demonstrate that the actual organization of cities is not sustainable but can be made so. By introducing not just the present but also the future, and its dangers, it conveys a moral imperative to do something. The successful metaphor not only describes but also prescribes. In its integrational function, it organizes not only meaning but also action across different discursive modes and their institutional and social contexts.

Metaphor both admits and conceals the imprecision of language use, with the consequence that words can be either insufficient in themselves, or can be used, unwittingly or deliberately, with different meanings, connotations and intentions. They are but one of the linguistic devices of discourse which call for a critical awareness with regard to the meaning of words and their use in everyday situations.

The environment as a field of knowledge

The environment has become a recognized field of knowledge to which many scientific disciplines and areas of technical expertise contribute. Positivist science presupposes the objective existence of the environment and the autonomy of the different specialist areas of knowledge that are brought to bear on it, which must then devise interdisciplinary practices and collaborative institutional arrangements to deal with it. Planning, as the coordination of knowledgeable practice on space-as-environment, assigns itself a key role in this respect, as we saw earlier.

Foucault developed a radically different approach to how fields of knowledge are constituted in his 'archaeological' studies, methodologically drawn together in *The Archaeology of Knowledge* (Foucault, 1989). Here he formalizes his idea of the constitutive nature of discourse, that is, that social reality is actively constituted through discourse in terms of the objects of knowledge, conceptual frameworks, social subjects and their relations, and so forth: and second, that discourses are interdependent, that is, that which was is written or said about one 'object' of discourse (political economy, medicine or the environment, for example) is related to and dependent upon that which is written and said about another, apparently independent object. Reaching these insights, however, involved a very particular approach to discourse. He argued that discourse should be conceived as an 'event' in order to capture the narrowness and singularity of each enunciation (or statement) and at the same time its location as a 'knot' in a network of other enunciations. This interdependency of enunciations constitutes what Foucault describes as a 'discursive formation'. This does not refer to the discourse (enunciation or text) itself but to the 'system of dispersion' or 'rules of coexistence' of diverse, unstable and often incompatible enunciations that constitute a field of knowledge. A discursive 'event' is thus both linguistic and social, singular to a particular historical time, place and set of circumstances.

What interested Foucault above all were the general rules governing what may and what may not be said, and the conditions of emergence of those rules. The 'rules of formation' of a discourse or set of statements were developed in relation to the 'formation of objects', 'enunciative modalities', 'concepts', and 'themes or argumentative strategies'. The general idea is that objects (for Foucault, fields of knowledge) are not determined by what is said about them in isolation but constituted through rules which are a function of other fields of knowledge, systems of institutional and social regulation of statements, the intertextual circulation of concepts or 'fields of statements', and their discursive organization into themes and theories. Foucault's own method was highly abstract, for example, preferring the term 'discursive formation' 'to avoid the use of words such as "science", "ideology", "theory" or "realms of objectivity", too impregnated with conditions and consequences', and concerned with the transformation of systems of knowledge over long periods of time.

Nevertheless, what may be drawn from this aspect of Foucault's work is the conception of fields of knowledge (such as the environment) being driven not by the environmental sciences but by a complex system of rules concerning the (discursive) relations between multiple fields of knowledge, and that the type of synthesis or discursive performance of urban environmental planning is not a product of planning's own specialist expertise but a function of its institutional position in relation to the discursive formation itself. Darrier (1999), for example, argues that Foucault's archaeological approach can be usefully employed to resist the 'fundamentalist temptation' of environmentalism. A further important point concerns the question of 'dispersion' or the idea that a discursive formation, unlike a scientific or professional discipline, makes no claim (there is no 'centre' to make it) to objectivity, internal coherence, logical consistency, and so forth. In this sense, the confusion and incoherence within the environmental field can be seen not as an epistemological weakness but a constitutive characteristic of environmentalism as a field of knowledge.

Foucault's archaeological approach leaves difficult questions open for applied discourse analysis. The elements of a discursive formation as analysed by Foucault – objects, enunciative modalities, concepts and thematic arguments – provide a broad framework for analysing a complex and open field of knowledge such as urban environmentalism. At the same time, it is clear that the strategy of such an analysis could not limit itself to what is being said about the urban environment alone and would have to explore how statements on it are conditioned by other fields of knowledge such as the urban economy, the geography of place, identity and citizenship, and so forth. Lanthier and Olivier's (1999) examination of the influence of legal and medical discourse on 'green' lifestyles is a useful illustration in this sense.

However, being concerned with the conditions of emergence and rules of formation of discourses, Foucault gave little attention to the analysis of texts and real instances of discursive practice that such an analysis implies; that is, of people actually saying and writing things. Fairclough (1992: 64) describes this as a problem of how to connect structure with practice or language in action, and his own solution is to argue a dialectic relationship in that 'discourse contributes to the constitution of all those dimensions of social structure which directly or indirectly shape and constrain it; its own norms and conventions, as well as the relations, identities and institutions which lie behind them'. This both sharpens the analytic focus and avoids the 'over-determination' of structure discernible in Foucault's early work, by seeing discourse as being enacted in an already constituted social reality (of both discursive and non-discursive practices) which simultaneously confirm and contest the rules of discourse.

Environmental knowledge/power

A far more influential aspect of Foucault's work in the urban and environmental fields has been his 'genealogical' studies which focus on the relationship between systems of knowledge and power. What, he asked, are the systems of power or 'regimes of truth' which produce and sustain certain systems of knowledge (Rabinow, 1984). Rejecting the notion of power as something to be possessed and imposed forcefully from above, Foucault argued that power operates 'from below' and incorporates people in more subtle process of domination. His famous studies on the hospital, the asylum and the prison demonstrate how such institutions became centres for new 'micro-techniques' of power, which Foucault describes in terms of examination and confession (Foucault, 1979; 1981). These techniques are processes of knowledge gathering and the organization of that knowledge through the social sciences. But they are also means for the exercise of power, in the sense of the institutionalized control over the production of knowledge and its rules of appropriation, of what can be legitimately said, by whom and in what circumstances. The resulting 'disciplinary' society is thus simultaneously a normalizing one, constituting and moulding individual subjects to perform the tasks socially required of them through the institutionally controlled discursive practices of criminality, medicine, education, sanity, and so forth.

Designed for controlling the masses, the disciplinary society operates in a highly individualizing way by extracting detailed information from the individual subject. In contrast to the exercise of sovereign power which visibilizes the sovereign, argues Foucault, the techniques of power/knowledge visibilize the ordinary individual through permanent scrutiny and evaluation, documenting and recording his/her traits and deviances, so that each individual becomes a 'case' in itself and 'an object for a branch of knowledge and a hold for a branch of power'. Power/knowledge thus objectifies people through examination and subjectifies them through the techniques of confession (the admission of weakness), which reveal and confirm people's inadequacy and draw them further into the domain of power (Fairclough, 1992). Educational, medical and professional examination (all now continued far into adult life), constantly control and expand the field of performance monitoring and measurement, whilst confession and counselling have expanded into our weaknesses as psychological individuals, partners and lovers, consumers, financial managers or with regard to the environmental friendliness of our lifestyles. It is against this pernicious exercise of power through everyday practices that Foucault called for 'resistance'.

Foucault's notion of 'regimes of truth' places considerable importance on institutions, thereby providing a firm basis for examining the politics of urban planning and the environment. Pioneering research in this direction is Hajer's (1995) well-known study of environmental policy-making which he termed ecological modernization. Hajer's proposition is that policy-making is the dominant way in which modern societies regulate latent social conflict and that the problems which policy addresses do not exist on their own, but are socially constructed so as to be manageable: the nature of the problems, the types of solution, and the institutional arrangements required to implement them, are dependent upon their construction through discursive practice. This discursive practice is fragmented and contradictory, realized by multiple actors using distinct cognitive bases and with often conflicting interests. How, he asks, can some seemingly coherent and generally acceptable account of problems and policy solutions be construed from such a confusing diversity of knowledge and interests?

Drawing additionally on the work of Harré and others, Hajer uses 'story-lines' and 'discourse coalitions' as middle-range analytic categories. He describes story-lines as 'narratives of social reality through which elements from many different domains are combined and that provide actors with a set of symbolic references that suggest a common understanding'. As such, story-lines are metaphorical devices which give general accounts of events to which actors are willing to subscribe in principle; they are something which 'sounds right'. They don't have to be internally coherent or scientifically rigorous but benefit from ritualization through repetition and create possibilities for problem enclosure. In other words, writes Hajer, 'a story-line provides the narrative that allows the scientist, environmentalist, politician, or whoever, to illustrate where his or her work fits into the jigsaw' by clustering knowledge, institutions and actors. Discourse coalitions are formed by actors around story-lines in order to maintain the dominance or hegemony of that general account. Subscription to a story-line has two major implications. First, it implies involvement in the social practices in which this discursive activity occurs: it involves a political commitment. Second, it involves an epistemological commitment, a tentative agreement on the nature and causes of a given phenomenon. Thereafter, 'the reproduction of a discursive order is then found in the routinization of the cognitive commitments that are implicit in these story-lines' (Hajer, 1995: 65). From a similar perspective, Sandercock (2003) has argued not only the importance of stories for a greater creative and critical understanding of urban processes (especially in multi-cultural situations) but that planning itself can be understood as 'performed story'.

Centring attention on institutions and coalitions deflects the difficult question of agency. Foucault is often criticized for minimizing the active subject and human agency by arguing the autonomy of discourse formations and the primacy of the rules of discourse over individual thought. It is contested that people are not meekly constituted through discourse, but use it consciously to form their own identities and political and cultural projects. Focusing on institutions allows such debate to be side-stepped. Institutions provide the sites around which social structures and individual subjects connect, and where policy dialogue is undertaken within discursively controlled limits but leaving open, hypothetically at least, the possibility of significant argument and contestation. Research in the field of urban policy, concerned as it is with the political and institutional frameworks within which planning operates, has taken up the opportunity offered by discourse analysis to reconsider the implications of new institutional landscapes for planning practice (Vigar *et al.*, 2000; Rydin, 2003).

Green governmentality

A further Foucauldian notion deriving from his genealogical studies is that of governmentality or the exercise of power through the control of populations. In sixteenth- to eighteenth-century Europe, argued Foucault (1991), government passed from the hands of the sovereign ruler to the state; with that transition, the finality of power ceased to be the maintenance of the sovereign him/herself and but rather the 'proper disposition of things' for increasing production and wealth. The security and the defence of sovereign territory continued to be important, but the exercise of power began to be primarily concerned with the populations that inhabited them. In this sense Foucault describes the defining trait of governmentality as 'bio-power' or the historical shift in the art of government to the management of the health, wealth and morality of subjects. This had two basic forms involving disciplinary action on the body ('anatomo-politics' of the human body) and the regulatory control of populations (or the 'species body'). The first centred on the body as machine, 'its disciplining, the optimization of its capabilities, the extortion of its forces, the parallel increase in its usefulness and its docility . . .'; the second focused on populations and their 'propagation, births and mortality, the level of health, life expectancy and longevity, with all the conditions that can cause these to vary'.

Foucault describes this as the 'entry of life into history'. Whereas the exercise of sovereign power was the right to take life, governmentality was the obligation to protect and nurture it. This, he observes, was indispensable to the rise of capitalism 'which would not have been possible without the controlled insertion into the machinery of production and the adjustment of the phenomena of population to economic processes' (Foucault, 1981: 140). The institutions of the state as *institutions* of power ensured the maintenance of the relations of production whilst biopower, as *techniques* of power, operated in the sphere of the social body and economic processes, meaning nothing less than 'the adjustment of the accumulation of men to that of capital'.

The idea of the focus of government on life holds obvious attractions for environmental critique. As Darrier (1999) observes in one of the first comprehensive examinations of the relevance of Foucault to environmental thought, Foucault rarely mentioned the environment in his intellectual work, had scant personal empathy with nature and was averse to any kind of foundational category such as 'nature' to which environmentalism generally adheres. However, not only does Foucault's work challenge environmentalism to examine its own conditions of emergence as discourse, his arguments concerning biopower and the control of human populations can be extended to include ecological concerns, with environmental management conceived as an extension of biopolitics to all life-forms and part of the normalizing strategy of ecological modernization at a global scale. This idea is developed through the notion of 'ecological governmentality', or the manner in which all biological life has become the object of scientific knowledge, calculation and strategy (sustainable development), a state of preoccupation and an ethical/normalizing principle for the conduct of organizational and individual behaviour.

Similarly, it is tempting to re-examine planning as a form of governmentality and try to 'trace its connections to the normalizing discourses that seek to render subjects and the spaces constituted through them as both manageable and free' (Huxley, 2002: 145). Some of the most obvious points of analysis would be land use zoning and regulation which formalize the classification, design, access to and control over differentiated space and pre-configure the types of social relation and interaction which are therein heralded as 'normal'. More particularly, neoliberal urbanization can be seen as a systematic redefinition of the normal, the necessity to re-draw the lines of 'proper' spatial organization and the contours of appropriate social behaviour. In this way, for example, urban planning can be seen as collaborating in the shift of the meaning of citizenship 'to a novel concept of the individualized subject responsible for his or her own well-being, supported largely through the market place, market orientation, clientelism, consumer fees, voluntarism, and criminalization of marginal behaviours and spaces', as Keil (2002: 247) argues in his study of Toronto. What is at stake, suggests Keil, is that urban development paraded on the neoliberal promise of opportunity, freedom of choice and freedom from government conceal the wilful subjection of people to ethical laws and norms that demand sacrifice and the redisciplining of populations.

Understanding cities as urban environments, centres of both the economy and populations and held to be vital nodes in the shift towards sustainable development, suggests the coming together of planning and green governmentality through urban environmentalism: a new disciplinary régime of neoliberal urban governance. Undoubtedly, neoliberal planning became caught up in the re-drawing of the boundaries between public responsibility and private duty, the citizen as client and customer and 'autonomous market participants who are responsible for their own success, health and happiness' (Isin, 1998, cited in Keil, 2002: 234). Similarly, the flexibilization and deregulation of land use and development controls can be seen to have been compensated by the disciplinary tactics of sustainable city management involving detailed control over lifestyle and the legal and moral obligations which underlie and enforce it. The urban environmental agenda of prudent resource use, responsible waste management, care of green space and preventive health can be seen to constitute a new régime of values and performance requirements for contemporary urban citizens, constantly examined through the techniques of remote surveillance, the production of local participatory environmental indicators, discriminatory domestic service charges and incentives, and so on.

It matters little that these rules may or may not be adhered to or have any objective ecological significance for the sustainability of cities. As Smart (1983: 115) observes, a disciplinary society should not be confused with a disciplined society, or the non-correspondence between government programmes and actual practices. The important thing from a Foucauldian perspective is the establishment of a subtle and complex field of social control over subjectivities, identities, duties and conditions to the full rights of citizenship. In this way, ecological strategies are tactics of governmentality which complement the declining disciplinary régimes of the bureaucratic welfare state in the individualized contemporary social order and form part of a 'new pluralization of social technologies' (Rose, 1996).

Discursive and non-discursive practices

The examination of discourse undertaken so far has passed from the field of linguistics through to questions of knowledge, power and techniques of government. The question now arises as to the relation between these different instances and effects of discourse and other 'moments' in the process of social change – between discursive and non-discursive practices – and how this complex picture might be operationalized in terms of the analysis of urban environmentalism. As has already been mentioned in passing, this is a complex matter involving questions of social structure and human agency as well as the role of discourse itself. There is nothing new as such about the general proposition that discursive and non-discursive practices influence each other. However, Foucault's general view was that systems of knowledge are organized in such a way as to have a certain autonomy and constitute relations of power, so that discourse 'is not simply that which translates struggles or systems of domination, but the thing for which and by which there is struggle, discourse is the power which is to be seized'. He emphasized the importance of the historical conditions of emergence of systems of knowledge and the régimes of power that produce, sustain and transform them, without proposing a general theory of social change as such.

The general nature of his theorizing and his reluctance to address political change much beyond the concept of 'resistance' was to cause considerable frustration but left many tantalizing openings for concrete enquiry. Within the field of urban studies, Harvey (1996) is one leading figure who seized the challenge, integrating discourse into his Marxist approach to historical–geographical transformations. After reminding us of the importance Marx attached to discourse, as made clear through a dialectical reading of Marx's work (for example, in Marx's conviction that is in the realm of discourses and beliefs that we become conscious of political issues and 'fight them out'), Harvey rejects a discursive determinism in Foucault, and interprets his 'general argument' in the following terms:

> The moment of discourse gains its seemingly autonomous disciplinary powers with respect to social life to the degree that there is amnesia with respect to the processes that both form and reinforce it. The function of discourse is to create 'truths' that are in effect 'effects of truth' within the discourse rather than the universal truths they claim to be. Such 'effects of truth' become particularly pernicious, in Foucault's view, precisely because they emanate from institutions (the asylum, the hospital, the prison) which operate as incarnations of power. His main aim is to undermine the 'effects of truth' and to show how truth in discourse is always an internalized effect of other moments in the social process ... Like Marx, he accepts that we are always confined to the world of material effects, though he broadens the notion of materiality to incorporate past discourses.
>
> (Harvey, 1996: 95)

Within a dialectical mode of analysis, Harvey (op. cit.: 78) schematically defines discourse as one of six 'moments' of the social process, using the term moment 'to avoid, as far as possible, any sense of prior crystallization of processual activities into "permanences"– things, entities, clearly bounded domains, or systems', and defined these moments in terms of:

- material practices, or the sensuous and experiential nexus from which all primary knowledge about the world ultimately derives;
- social relations, or the forms of sociality that human beings engage in and the more or less durable ordering of social relations to which this sociality may give rise;
- institution building, or the organization of political and social relations between individuals on a more or less durable basis;
- the 'imaginary', or the thoughts, fantasies and desires about how the world is (ontologies), how better understandings of the world might be achieved (epistemologies), and how I/we want to be in the world;
- power, its internal heterogeneity, unequal distribution, and role with regard to the ordering of society;
- language/discourse, or the vast panoply of coded ways of talking about, writing about and representing the world.

These six 'moments' bear a striking similarity to the set of relations with which Foucault was concerned, but proportion a more applicable framework for analysing urban social change in the city. In effect, discourse ceases to be the centre of attention to become a key ingredient in the dialectics of social change. Discourse, then, neither mechanistically carries power nor directly brings about change. The general approach of dialectics holds that each moment internalizes the others in a complex and often ambiguous way. In rejecting the idea of a neat, circular causality of events,

Harvey makes the point that in the process of translation from one moment to another, 'slippage' occurs so that translation is always a metamorphosis rather than mimesis. As a result, there is never a perfect coherence between different moments, but rather a general correspondence characterized by a heterogeneity which derives from the conflicting effects from all other moments. Moreover, Harvey's Marxist approach places capital accumulation at the centre of the tensions and contradictions between and within these 'moments' in order to engage the more subtle fields of contradiction which modern capitalism has thrown into being, including the environmental 'crisis' and responses to it through urban management practices.

The environment as urban ideology

The theoretical explorations of the preceding sections help develop our understanding of how the propositions and effects of urban environmentalism are achieved through discourse and in an ideological fashion in the sense of naturalizing social change and the power relations it conceals. Nevertheless, the discourse approach to the analysis of social change tends to elide the question of ideology. The notion of governmentality or the exercise of power through the discursive control of everyday social practices bears a strong resemblance to how ideology operates, in the Althusserian sense of the constitution or 'interpellation' of individuals as social subjects. However, Foucault himself was reluctant to use the term ideology for several reasons: its association with truth and falseness (he was more concerned with seeing historically how effects of truth are produced within discourses which themselves are neither true nor false), its presupposition of a unified subject (rather than the 'decentred' subject constituted by multiple and contradictory discursive and non-discursive practices), and his concern that ideology is usually placed 'in a secondary position relative to something which functions as its infrastructure, as its material, economic determinant', which he himself replaced by discourse (Foucault, 1980; Howarth, 2002: 117).

Nonetheless, if discourse is constitutive of power and power is everywhere, it is 'tolerable only on the condition that it masks a substantial part of itself. Its success is proportional to its ability to hide its own mechanisms' (Foucault, 1981: 86). This concealment and naturalizing of power is a basic characteristic of ideology. As Fairclough (1992:. 67) argues, 'different types of discourse in different social domains and institutional settings may come to be politically or ideologically "invested" (Frow, 1985) in particular ways', although Fairclough himself prefers the concept of hegemony, with its emphasis on 'leadership as much as domination across the economic, political, cultural and ideological domains of a society'; a never-ending quest to maintain power as a whole. Fairclough argues that discursive practices contribute to and form part of the struggle to maintain hegemony in the sense of channelling the boundaries within which conceptions of the world are materialized in institutional practices and common sense understandings of everyday life. However, ideology can be seen as having stronger connections to the sort of phenomena associated with urban environmentalism during a process of the transformation of cities through the restructuring of capital, and in the following sections we explore how the environmental conception of space acquires this quality. To do this we first make a brief theoretical incursion into the notions of space and ideology.

The production of space and the environment

The historical nature of the environmental 'crisis' is without dispute. What often tends to be taken for granted is the spatial nature of the environment. The notion of sustainable development has contributed to the imposition of the temporal dimension over the spatial one, notwithstanding the fact that environmental problems have a self-evident spatial manifestation, such as where to locate a waste disposal site, the location of pollution sources and geographical dispersal patterns, the coordinates and relations of ecosystems and global climate patterns and so forth, and the management of environmental problems always contains a spatial strategy. However, whilst space is used to delimit the field of distribution of environmental systems and particular problems, it is less frequently employed as a means of understanding the problem itself; space disappears, as it were, into the environment.

The theoretical resources of urban studies provide a valuable contribution to a more complete understanding the environment, in particular the body of critical theory concerning the social construction of space (Harvey, 1982; Smith, 1984; Castree, 1997). The conception of space developed by Marxist urban sociologists in the 1970s and 1980s argued that space is socially constructed not just in the architectural sense (the 'built environment') or simply as a concentrated form for the production and circulation of economic value (the 'economic environment') or merely as a reflection of the cultural values (the 'cultural environment') through which society can be interpreted. They argued that space is constitutive of society. For example, Castells (1983) insisted that '[S]pace is not a "reflection" of society, it *is* society', being one of society's fundamental material dimensions and constitutive of a mode of production; spatial forms are produced, as all other objects, by human action, express power relationships and the interests of the dominant class. In a similar vein, Harvey (1989) argued that rather than directly approaching the city as a physical artefact or legal-political entity, it should be understood in terms of the social dynamic marked by processes of 'capital circulation; the shifting flows of labour power, commodities and capital; the spatial organization of production and the transformation of space-time relations, movements of information, geopolitical conflicts between territorially-based class alliances, and so on'.

We argue that a geographical–historical materialist approach to space provides a valuable route to understanding the environmental problematic and for developing a useful critique of urban environmentalism. Its key contribution consists of conceptually putting the environment in its place. Instead of space disappearing into the environment, the environment can be seen as absorbed within space and spatial production. In the process, the objectivization of natural space is negated, and replaced by a dialectic which relates and integrates the environment into the general movement of society and the urbanization process in particular.

The theoretical foundations and implications of this proposition deserve further clarification, and to this end we turn to the work of Henri Lefebvre, and especially *The Survival of Capitalism* (1976), perhaps his clearest and most direct account of capitalist space. For Lefebvre, a Marxist political philosopher and activist who, after 1968, became increasingly interested in the urbanization process, the central question was not how capitalism could be overthrown or transcended, but how it survived. He became convinced that this survival was through space:

> The concept of the production and reproduction of social relations resolves a contradiction in Marx's thought which, to him, could not have appeared as a contradiction. Marx thought that the productive forces constantly flung themselves against the restrictive limits of the existing relations of production (of the capitalist mode of production), and that revolution was going to leap over these constraints. Partial crisis was going to change into a general crisis . . . But what has happened is that capitalism has found itself able to attenuate (if not resolve) its internal contradictions for a century, and consequently, in the hundred years since the writing of *Capital*, it has succeeded in achieving 'growth'. We cannot calculate at what price, but we do know by what means: by occupying space, by producing space.
>
> (Lefebvre, 1976: 21)

For Lefebvre, the urban is the locus of power and centre of financial, political and information flows, around which the contradictions of capitalist production were attenuated. He saw the internal contradictions of capital-labour conflicts as centred within the urban itself, whilst the peripheries of both a social (the unemployed, youths, women, immigrants) and spatial (the rural, the regions) kind were ordered from the urban centres in accordance with the amplified requirements of accumulation or 'enlarged reproduction'. In contrast to Castells and many other Marxist urban sociologists of the time, who where more concerned with the production process and the role of the city and its infrastructure (or means of collective consumption) in that process, Lefebvre privileged the concept of reproduction. He argued that it is the productive process which introduces that 'most general of products' – space – into social existence, and in so doing confers space with the conflicts proper to the contradictions of the production process of capitalism itself. However, the differentiated and conflictive nature of socially produced capitalist space acquired a deeper significance for Lefebvre:

> The centre–periphery relation is neither the sole nor the essential conflictive relation, in spite of its importance. It is subordinate to a deeper conflictive relation between, on the one hand, the *fragmentation* of space (its *practical* fragmentation, since space has become a commodity that is bought and sold, chopped up into lots and parcels; but also its *theoretical* fragmentation, since it is carved up by scientific specialisation), and, on the other hand, the global capacity of the productive forces and of scientific knowledge to produce spaces on a planetary and even interplanetary scale.

> *This dialectised, conflictive space is where the reproduction of the relations of production is achieved. It is this space that produces reproduction, by introducing into it its multiple contradictions, whether or not these latter have sprung from historical time.*
>
> (Lefebvre, 1976: 19, italics in original)

In this context, conflict arises at the point in the which the 'distant order' of the abstract space of the relations of production at a global scale 'brutally invade' the 'local order' of the relations of production at the scale of the community, neighbourhood or city. Lefebvre describes this conflict as the production of difference at the juncture between the logic (capitalist, abstract) and dialectic (real, political) of space. This is a continuous process which becomes more self-evident in times of systemic change, as illustrated by the radical urban transformation required by the globalization of capital and executed through neoliberal urban development policy.

In this way, Lefebvre argues, space becomes the privileged sphere of the reproduction of the social relations of production, rather than the economic relation contained within the unit of production or the economic sphere in general. The social relations of production are 'reproduced in the market in its widest sense – in everyday life, in the family, in the town'; space is where the reproduction of the relations of production is *achieved*. In turn, the achievement of social relations in space requires the power to control space. This, Lefebvre maintains, is a question of strategy, or 'ensembles' of mechanisms such as the legal system and scientific knowledge (discourses) which reproduce through space an everyday consciousness concealing the abstract relations of production, with the specific end of making conflict appear as somehow coherent and natural. The sciences of space and particularly planning, argues Lefebvre, have precisely this function of providing a rational explanation of a conflictive spatial order and making it appear as a coherent whole.

Lefebvre singled out systems theorists in particular as the ideologues of capitalist space, in their pretence to enclose an open-ended reality and provide a sense of order, inclusion and control. In the intervening years since Lefebvre's work, have ecosystems theorists and the urban experts who build on them become the new ideologues of capital? Despite inclusion having now become an explicit goal of social policy, is the environment a more subtle, ideological way of achieving a sense of social cohesion? That the control of space is ideological is, of course, central to Lefebvre's work, and his views on ideology and the environment will be examined later. However, we can already draw out the general direction of Lefebvre's critical stance with regard to the environment: its inference of naturality, its capacity to rationalize space and the implicit systems character of this ecological version of space with its pretension to wholeness and control; that is to say, its ideological potential in postmodern cultural order of flexible accumulation. In *The Production of Space*, Lefebvre connects the political significance of space to the environment in the following terms:

> Some of the new contradictions generated by the extension of capitalism to space have given rise to quickly popularised *representations*. These divert and evade the problems

involved (i.e. the problematic of space), and in fact serve to mask the contradictions that have brought them into being. The issue of pollution is a case in point. Pollution has always existed, in that human groups, settled in villages and towns, have always discharged wastes and refuse into their natural surroundings; but the symbiosis – in the sense of exchange of energies and materials – between nature and society has recently undergone modification, doubtless to the point of rupture. This is what a word such as 'pollution' at once acknowledges and conceals by metaphorizing such ordinary things as household rubbish and smoking chimneys. In the case of the 'environment', we are confronted by a typically metonymic manoeuvre, for the term takes us from the part – a fragment of space more or less fully occupied by objects and signs, functions and structures – to the whole, which is empty and defined as a neutral and passive 'medium'. If we ask, 'whose environment?', or 'the environment for what?', no pertinent answer is forthcoming.

(Lefebvre, 1991: 326, italics in original)

This metonymic manoeuvre is something we will come back to later. Before that, however, a closer look needs to be taken at the implications of the social production of space for that which, increasingly confused nowadays with the 'environment', still frequently goes under the name of 'nature'.

The ideologization of nature

It is a common enough theme of cultural anthropology that in pre-capitalist societies, nature provided the principal framework for the spatial and temporal organization of social activity. The natural rhythms of space and time, the appropriation of nature's products and forces, their endowment with meaning as super-natural phenomena, the use and symbolic values of natural space and so on, acted as the parameters within which everyday life unfolded. This natural organization of space holds an obvious fascination for certain currents of environmental thought which, drawing from anthropology, are able to elaborate a comparative demonstration of the 'waywardness' of modern Westernized culture. The much more significant point, however, is that the social production of capitalist space implies the creation of a different structuring principle, that of surplus value, and it is to this, and its implications for society's relation to nature, that we now turn.

Lefebvre's (1976) assertion that in the process of the social production of space, natural space is destroyed and material space transformed into a social product, is probably one that can be widely accepted in its generality, but less so in its explicative function or political consequences. After all, the processes historically associated with capitalism, such as industrialization and the increasingly technological exploitation of nature, provide ample evidence of the destructive transformation of space under capitalist development. Similarly, Lefebvre's proposition that the growth of productive forces creates new (environmental) contradictions is at the heart of the contemporary debate over sustainable development.

However, the underlying proposition of Lefebvre is not that nature is transformed, but that *space* is transformed, in the sense that it acquires a radically different form: a social one. It is at this point that difficulties arise for orthodox environmentalism, since the most generally used conceptual framework for understanding the environment is that of a relation, or more specifically, the evolution over time of a general relation between 'society' and 'nature', mediated by technological development. In the more sophisticated versions of this conceptual approach such as constructivism, mediation occurs through symbolic form, including technology but also belief systems, values, institutions and so forth; in other words, the relation is stated in terms of 'culture' and 'ecosystems'. Even so, the question of space tends to be left in the background, condemned as the ahistoric terrain on which this relation unfolds. Lefebvre proposed bringing space to the forefront and, in so doing, redefined the conceptual ground for the examination of problematized nature.

In this context, the critical question concerns not the transformation of natural space, but what happens to 'nature' in the creation of a new space, a socially produced one. In earlier sections we saw how a constructivist approach argues that there is no single 'nature' but, rather, multiple and contested 'natures' constructed according to diverse social and cultural practices. This is an important contribution to understanding contemporary environmental problems and politics, without necessarily addressing the broader and spatially defined question posed above. To this effect it is useful to examine Schmidt's (1971) account of Marx's view of nature which provides a key and incisive insight: that nature is necessarily ideological in socially-produced capitalist space. The briefest of outlines of Schmidt's argument is presented as follows.

Marx's general position was that mankind, in its biological and anthropological generality, is irremediably conditioned by nature. Nature in its totality – humankind and the rest of nature – constitute reality; that which distinguishes humankind from the rest of nature is its conscious, practical, transforming action. In his later analytical works, Marx elaborated on the historical character of this relation, and argued that in capitalist society it is the nature of relations between human beings which is the key to understanding the relation between humankind and nature, of which humankind is a part.

In *The Economic and Philosophic Manuscripts*, Marx had already set out the idea of alienated labour which 'separates man from the object of his production, and therefore from his "species being", his objective reality as a member of the species, which transforms his advantage over animals into the disadvantage that his "inorganic body" (nature), is confiscated from him' (Marx, 1988). In other words, life itself is presented as merely a means to life. In his later works, Marx expanded on this theme. He showed how, under the capitalist mode of production, nature becomes subordinated to society, not on the basis of some human delusion or divine will, but through concrete social practices. As Schmidt observes, Marx argued that from tribal to feudal society, individuals interact on the basis of their *natural* relations (family, clan, tribe), and interact with nature on the basis of a *natural* division of labour (according to age and gender) and undertake work within the presuppositions of the *natural* conditions of existence. Social production consists of the production of use values, directly related to the reproduction of the individual and his immediate natural community, where surpluses are exchanged in the form of barter or tribute. Whether land ownership is private or collective is irrelevant.

With the emergence of capitalism, Marx argued, the domination of nature acquired a new quality. This quality, which is only fully realized in the relation between wage-labour and capital, involved a radical separation from nature, not in the sense of the technological development of the means of production, but in the sense that work is no longer subordinated to nature, but to capital. The capitalist mode of production removes the worker from nature in a double sense: from nature as a medium of work (thereby creating a free and dispossessed labourer able to freely sell his labour force), and in his/her condition as a natural being (now reduced to a depository of labour and a commodity). In the process, individual reproduction is achieved through the production of commodities or exchange value. It is not that just the products of intellectual labour, the service sector or informational society are immaterial and insufficient in their specialized forms to guarantee individual reproduction. Marx shows how physical commodities as well are immaterial, in the sense that they consist of use value plus exchange value, or 'value'; this commodity form of value is 'physically metaphysical', in its classical form the expression of the amount of labour invested in an object which 'cannot therefore contain natural material at all in the same way as, verbi gracia, the exchange rate can'.

It is thus what Marx called the labour process and the commodity form of wealth (however sophisticated they may now have become) that provide the key elements. On the one hand, as nature ceases to be the organizing principle of social relations, the labourer is reduced to a mere producer of exchange value; that is to say, as a worker he is totally determined by society, which implies the total negation of his natural existence. On the other hand, the immateriality of the commodity form means that it is in itself blind to nature and to the repercussions which it might have for nature. In consequence, nature, as much in its contemporary ecological conception as in earlier romanticist idealizations, can only appear in the plane of the 'superstructure', that is to say, in the realm of politics, ethics, aesthetics, law and the combined action of the state.

In short, both nature and the alienated natural subject have to be reconstructed ideologically. To borrow Schmidt's (1971) felicitous term, in the process of capitalist industrialization *nature 'solidifies' in the abstract*. Objectively relegated to the background of social activity, nature is rescued and concretized in the sphere of ideas. And the formation of these ideas on nature, however 'scientific' they may be, never ceases to originate in and express the needs of the reproduction of the general conditions of production – the reproduction of the means of production and of the relations of production.

Ideological forms of nature

The extension of capitalist relations across space and ever further into nature's materiality implies a progressive 'ideologization' of nature, an ungainly term but more exact than ideology, since it conveys the process of abstraction rather than any fixed notion of nature. The ideologization of nature is an objective consequence of capitalist production and an integral part of the reproduction of the relations of production. This is something which remains intrinsic to capitalism, irrespective of the enormous growth in the means of production (the technology and management of production), innovations in the organization of production and changing conditions of labour (flexibilization), the seductive abundance of goods and services (and the emergence of the so-called consumer society) or transformations in the régime of production (informational mode of production). In fact, the extension of capital into ever greater areas of personal life and the effects of generalized commodification on personal subjectivity have had the effect of sharpening the sense of estrangement from nature, at the same time as they frustrate at every turn individual efforts to reconnect with the natural conditions of existence. The growth in ecological tourism, its inexorable impoverishment in the form of the packaged spectacle and the tourist gaze (Urry, 1990) provides but one illustration of spatial significance.

The significant point to be made is that the ideologization of nature is, simultaneously, the creation of an ideological form of space, since nature is itself the natural form of space. As a consequence, as Harvey (1996: 119) concludes from a slightly different line of argumentation, '. . . all proposals concerning "the environment" are necessarily and simultaneously proposals for social change and action on them always entails the instantiation in "nature" of a certain régime of values'. The environment itself is, of course, a relatively recent term and a value-laden conception of nature. From a wider historical perspective, it might be argued that the need to restore value to nature is surely one of the defining characteristics of the modern capitalist era, an issue which has been given considerable attention through urban planning and design.

For the most part, the search for value in nature could be consigned to the realm of bourgeois academic philosophy, art and the aesthetics of everyday life of the leisured classes. The eighteenth-century saw the development of an existential sensibility towards the tragic beauty of nature (Todd, 1986), to be replaced in the nineteenth-century by romanticism and a more serene, scientifically-informed contemplation of the wonder of nature. The effects of these changing historical attitudes towards nature on urban form have been explored, for example, by Green (1990), and in the Latin American case, Arango (1980). As the twentieth century wore on, ideas concerning the place of nature in urban development were important in early planning thought from Ebenezer Howard to Le Corbusier and Frank Lloyd Wright. By the last quarter of this century however, utopic romanticism of this modern kind became increasingly problematic (Brand, 1996). The environmental 'crisis' made transformed nature into something which could no longer be either contemplated or spatially manipulated in the comfort of remove. It came to impinge on everyday life in a relentlessly threatening way, to the degree that the dominant relation with nature has become that of scientific management and a moralizing mode of interpretation. As nature became a problematic spatial experience, its representation had to be radically transformed. Nature acquired a more serious aspect: it became an ontological problem and a practical one.

There are two peculiarly postmodern responses to the contemporary spatial problematic of nature. One is based on an in many ways alarming increase in scientific-technological capacity. Within this line of thinking, the deterritorialization of social activity through developments in communications and information technology is linked with the reconstitution of nature through genetics and biotechnology. Haraway's (1991) influential thesis on the 'reinvention of nature' argues that science and discourse are producing a new entity, the cyborg, a hybrid creature which is neither organism nor machine, but rather a 'humachine' or strategic ensemble of organic, textual and technical components. The result is, so the argument goes, the advent of cyberculture involving a radical change in biological and social life. Inquiry of this sort (see for example, Luke, 1997) is necessarily speculative, and tends to consciously conflate science and fiction. However, implicit in this vision of the reinvention of nature is the reinvention of space as well. Cyberculture is inconceivable without its correlative construct: virtual space.

A second response is much more familiar, and in a sense is the opposing extreme of practicality. Sustainability discourse is based on the ecological–scientific understanding of nature 'as it exists', through which nature is dissected up and reconstructed as a system to be managed. Here nature does not disappear in technological reinvention, but rather reappears as 'environment' (Sachs, 1992). Moreover, with regard to the environment as the spatialized conception of nature under global capitalism, Escobar observes:

> As the term is used today, 'environment' includes a view of nature from the perspective of the urban-industrial system. Everything that is relevant to the functioning of this system becomes part of the environment. The active principle of this conceptualisation is the human agent and his/her creations, while nature is confined to a passive role. What circulates are raw materials, industrial products, toxic wastes, 'resources'; nature is reduced to a stasis, a mere appendage to the environment. Along with the physical deterioration of nature, we are witnessing its symbolic death. That which moves, creates, inspires – that is, the organising principle of life – now resides in the environment.
>
> (Escobar, 1996: 52)

It is, perhaps, this spatializing conception of nature, more than the rationalizing strategy of sustainability, which secures the disciplining effect of the environment on social activity, since the notion of the environment serves to bring the 'rational' exploitation of nature into everyday spatial experience. Whatever the case, as Lefebvre (1976: 27) observed three decades ago, everything conspires against nature and its authenticity. The environment, with its focus on systemized, managed nature and its problems such as pollution, does nothing to solve the problem:

> 'The central problem is not "the environment" but the problem of space. An ecosystem, once broken up, cannot reconstruct itself. Once even a fragment disappears, then theoretical thought and social practice have to recreate a totality. This cannot be done in bits and pieces; therefore they have to produce a space'.

The environmental sciences, which now so dominate our conception of space, pretend to protect and recreate ecosystems. Through the knowledge bases of the natural sciences and the totalizing schema of ecosystems theory, they purport to reproduce stable natural dynamics in relative isolation; they pretend to reproduce the natural organization of space, whilst ignoring the totality of space – and therefore the social nature of environmental design – by reducing human activity to a simple component or control system. The real effects of environmental science are not so much to recreate ecosystems as to reproduce social space. In short, the modern capitalist ideologization of nature has assumed its postmodern ideological and peculiarly spatial form. Before exploring the implications of this, however, it is important to clarify the nature of ideology.

On the nature of ideology

To recapitulate, the argument so far maintains that under capitalism, space assumes a particular form: not only occupied, appropriated and given meaning, but socially constructed in the unique sense of being wrested from its natural organization through the creation of abstract (exchange) value. Exchange value is, at the same time, the medium which facilitates the expansion of capital over space and into ever wider areas of social life, in an unlimited process of commodification, including the commodification of nature. This double movement of abstraction and commodification results in the ideologization of nature in the very process of the spatialization of capital, the contemporary form of which is 'environment'. At this point, having established that the environment is the dominant ideological form of space/nature in late capitalism, more detailed consideration needs to be given to the meaning of the term 'ideological'. Given that ideology is almost as open a term as 'space' or 'environment', clarification is required with regard to the specific interpretation employed here and modes of operation in relation to urban space.

As Lefebvre (1976) comments, the concept of ideology 'has been extended beyond all measure, and this has sterilized it'. It is popularly used to refer to any body of beliefs, and thus can be used to refer to the expression of any opinion or set of ideas. This is what Thompson (1990) describes as neutral conceptions of ideology, in the sense that they 'purport to characterise phenomena as ideology or ideological without implying that these phenomena are necessarily misleading, illusory or aligned with the interests of any particular group'. Often popularly associated with a world-view, ideology also has a long and complex history in contemporary modern philosophy. Eagleton (1991) points out that the possibility of ideology arose within the growing tensions in the Enlightenment between reason and superstition, empiricism and metaphysics. It has always been and remains a contested domain of social theory, and as Eagleton observes, in the cultural climate of postmodernity was once again questioned as a relevant concept, whether by Habermas' relegation of it to technocracy and the abandonment of meaning, its replacement by irony in the works of Rorty, or as simply being redundant in the complexity and incoherence of contemporary social life.

The materialist contribution of Marx was to crystallize those Enlightenment tensions around epistemic and political issues, the former related to questions of truth/consciousness, the latter to questions of power. Moreover, of course, he rooted ideology in the capitalist mode of production, with its mystificatory role in the social presentation of the real relations of production. The general questions addressed by ideology – consciousness, power and domination – lose their ideological significance if detached from those real conditions of existence or the materiality of social life. Thus, for Lefebvre, the important category is neither consciousness nor power, but that of practice. In this sense, ideology is not so much a set of ideas but the effect of the employment of knowledge/power in such a way as to reproduce social relations. Lefebvre's (1976: 29) definition of ideology as 'any representation which contributes either immediately or "mediately" to the reproduction of the relations of production' will subsequently be adopted to guide the following discussion.

This definition introduces further key characteristics of ideology. First, ideology is not a thing, but rather the representation of a relation. This representation may take the form of a thing (discourse, institution, object, image, metaphor, idea, myth, etc.) but it is the function of that thing in relation to the maintenance of social relations which defines it as ideological. Second, and deriving from the first, ideological representation is not rigid and static, but a fluid affair due to its relational character within dynamic social processes. As Lefebvre (1976) points out, ideology has to be constantly reproduced as '[t]here is no reproduction of social relations without a certain production of those relations; there is no purely repetitive process'. Consequently, ideology masks the production of new relations as much as the renewal of the general relations of production. Third, ideological forms are not self-proclaiming justifications of a régime, but are as indirect and invisible as possible. The success of ideology depends on it not appearing as such. Ideology can and does profess itself to be non-ideological, but the most effective ideology is that which is closest to the social practice of capitalism, and hence provides the illusion of a natural reproduction of the relations of production.

The study of ideology is therefore a critique of the relations of production, a critical analysis of the uses of representations to maintain the social relations of production. It is this inherently critical stance which produces widespread resistance to the notion of ideology, since it necessarily involves a challenge to everyday 'common sense' or 'self-evident' knowledge, the sovereignty of the individual as a knowledgeable subject and his/her self-determination within a particular social system, at the same time as it exposes the material self-interest (position within social practice) upon which such sovereignty is founded.

We earlier mentioned the relation between discourse and ideology, from which it may be concluded along with Eagleton (1991) that ideology is less concerned with a particular set of discourses than a particular set of *effects* within discourses. Whilst discourse provides one powerful means of communicating knowledge in the exercise of power, the general turn in social critique towards discourse has itself been criticized from a materialist perspective for losing contact with the non-discursive realms of social activity and material reality. Eagleton, for example, makes the following typically penetrating observation:

> It is surely a little immodest of academics, professionally concerned with discourse as they are, to project their own preoccupations onto the world as a whole, in that ideology known as (post-)structuralism . . . The neo-Nietschean language of post-Marxism, for which there is little or nothing 'given' in reality, belongs to a period of political crisis – an era in which it could indeed appear that the traditional social interests of the working class had evaporated overnight, leaving you with your hegemonic forms and precious little material content. Post-Marxist discourse theorists may place a ban on the question of where ideas come from; but we can certainly turn this question back on themselves. For the whole theory is itself historically grounded in a particular phase of advanced capitalism, and it is thus living testimony in its very existence to that 'necessary' relation between forms of consciousness and social reality which it so vehemently denies. What is offered as a *universal* thesis about discourse, politics and interests, as so often with ideologies, is alert to everything but its own historical grounds of possibility.
>
> (Eagleton, 1991: 219–20) (italics in original)

From this summary discussion of ideology, two important aspects should be highlighted. First, ideology, in all its shifting forms, is above all a practical issue. As understood by Althusser (1969), it is concerned with producing forms of subjectivity which directly impact on the way people carry out their everyday lives, through the representation of the imaginary relationships of individuals to their real conditions of existence. Knowledge and power are ideological only to the degree that they contribute to that subjectivity or constitute control over those relations, whilst the question of truth may be held in suspense. This seems particularly pertinent to the environmental problematic, with its scientific uncertainty, the continual unearthing of new and contradictory evidence of all kinds, and the political controversy it inspires. Epistemological and empirical confusion as to the state of the environment as a biophysical entity is of little importance; in fact, confusion and controversy may be regarded as positive ideological features since they contribute to the fixing of attention on the environment and away from social relations, except in a carefully controlled manner.

Second, ideology is neither explicit statement nor invention unrelated to people's knowledge of their circumstances gained through experience or 'unmediated' reflection on their social condition. Ideology perforce deals with significant issues. However, it does fulfil the function of providing a sense of naturalness to those issues, whilst deflecting attention from the unjust, alienating, conflictive and unstable social relations of capitalism. The environment may be seen as having now become the canvas of that natural order, or inversely, its imminent destruction. Harmony and apocalypse, in and through the environment, cooperated with each other to circumscribe the critical political imagination of the late twentieth century. Again, it is space which is now the object of ideology and the legimating device of the state.

Conclusions

The theoretical explorations developed in this chapter have addressed the question of how to understand urban environmentalism as a social project with only tenuous foundations in the ecology, the natural sciences and the positivist logic of orthodox sustainable development. The initial focus on discourse sought to clarify how knowledge on the environment is constitutive of power to control the meanings and rules of participation in talking about and acting on the environment. In this sense institutions were seen as vital sites for organizing power/knowledge, particularly through the urban policy process but also in terms of the mobilization of meaning through concrete urban management practices, presented nominally as an objective response to external environmental problems but effectively acting as a means of regulating populations.

At the same time, discourse theory makes it quite clear that it is insufficient to focus on discourse on the environment alone. First, any discursive formation is open and regulated by the knowledge produced in other discursive fields. Nominally founded on the compartmentalized knowledge of the environmental sciences, discourse theory requires that understanding urban environmental discourse must take into account the influence of other urban discourses on themes such as competitiveness, appropriate forms of governance, the quality of cities and urban life, notions of citizenship, and so forth. Attention to discourse, we maintain, is a crucial way of developing a critical understanding of the relatedness of the environment to other urban issues and the political content and social implications of urban environmental agendas. Urban environmentalism can be seen as a form of green governmentality which structures and regulates ideas, attitudes and sensibilities in accordance with the demands of social change in general.

Second, discourse theory allows knowledge on the environment to be seen not as objective fact but socially-produced understanding, and that it is therefore necessary to consider its historical conditions of emergence. In the case of urban environmentalism, a full understanding requires its contextualization in the period of the globalization of capitalism, urban economic restructuring and neoliberal government, with all the radical transformations in the physical and social organization of cities that this implied. Discourse is seen as one 'moment' in this complex process of change, but a moment which at the same time internalizes all the others. In this sense, discourse analysis is implicitly the analysis of social change in general, but critical discourse analysis requires that what is being internalized in discourse be given explicit attention.

Urban environmental discourse is about nature and its significance for the organization, management and building of cities. The second part of this chapter explored the general concept of nature from the perspective of its construction not through discourse but through the production of capitalist space. It was argued that natural space is no longer the creative force and structuring principle of social organization, now replaced by surplus value, with the consequence that nature is 'solidified in the abstract' – in the realm of ideas – and necessarily acquires an ideological form. However, in the global space of advanced capitalism, nature and space are fused once again into a single, universal construct – the environment – but now under the strict aegis of a socially-produced rather than naturally occurring order. As the hegemonic idea of nature, the environment allows nature to be objectified and subjected to management procedures. As a consequence, urban environmental management, although ostensibly about nature, is necessarily concerned with the management of social space, with all its regulatory and disciplinary implications.

This chapter set out a general theoretical approach for the understanding of urban environmentalism and some methodological signposts for concrete analysis. Neither discourse theory nor a dialectical approach to social change hold that there is some kind of direct and immutable relation between what is being said in the name of the environment and concrete processes of urban change. Rather, the relation between discourse and change in general is one of internalizations, slippages, irregularities and incoherence, wherein the effects produced are more significant than the analytic and management structures which support them. Furthermore, given that discourses are produced in historically and geographically specific circumstances – in an already produced reality – then existing cultural understandings, residues of earlier meanings, institutional frameworks and urban trajectories can be expected to modulate what and how particular effects are achieved. The studies of urban environmentalism in action in the following chapters explore these kinds of phenomena in detail.

Chapter 5
Urban environmentalism in action 1:
competitiveness and the quality of life in Birmingham, UK

Introduction

The UK's second city is situated in the centre of England and dominates the West Midlands conurbation, an industrial region consisting of seven unitary local authorities. Birmingham currently accounts for half of the conurbation's total population of just over two million people and for over a hundred years its metal manufacturing economy drove both regional industry and led the British engineering and automobile sectors. It was in many ways the unfashionable hub of British urban life, with a solid civic tradition and manufacturing base which inspired more respect than admiration.

Birmingham is, above all, a creation of the industrial revolution and in the early nineteenth-century could claim that 'there is scarcely a town in America or Europe that is not indebted for some portion of its luxury or its comfort to the enterprise and ingenuity of the men of Birmingham' (cited in Briggs, 1968: 184). The wealth of Birmingham was based on engineering. As coal mining, iron working and metal manufacturing developed throughout the region during the nineteenth-century, Birmingham became the centre of technological development and skilled labour. The combination of an economy organized around small firms, harmonious class relations and strong political organization was the basis for innovative improvement of the haphazard order and filthy conditions typical of early industrial cities. From the mid-nineteenth century onwards, Birmingham led the way in the municipalization of services and slum clearance. By the late nineteenth century it had acquired an international reputation as being one the best-governed cities in the world. Recent rivalry with Manchester for second-city status echoes nineteenth century competition between them, when both cities began making legitimate claims to international importance, political enlightenment and municipal progressiveness.

Economic prosperity and urban growth continued into the twentieth century. In the post-war period Birmingham, eager to assert itself as a major pioneering centre, embraced modernist redevelopment with unusual enthusiasm. True to its engineering and automobile tradition – on the eve of collapse – and centre of the emerging national motorway network, Birmingham was rebuilt on the concrete aesthetic of urban expressways, tower blocks and grey shopping centres. But the 1960s also proved to be a turning point in the city's economic fortunes. Diminishing profits and technological stagnation in the engineering sector were temporarily offset by mergers and acquisitions, leading to increased industrial concentration and poor long-term prospects (Walker, 2000). This decline continued into the 1970s and currently the manufacturing sector (especially motor vehicle and component manufacturing and supply) accounts for less than 20 per cent of total employment in the city.

The past 20 years of urban development have in many ways been a single-minded attempt to rebuild the city's image and attack its economic dependency on manufacturing. To meet the challenge, Birmingham enjoyed considerable institutional resources but was faced with a daunting urban legacy of a decaying industrial fabric, abandoned and contaminated sites and an outmoded city image. Development of the financial, cultural and retailing sectors in recent years has underscored a more glamorous architecture and urban design in the city centre, as attention turned away from conserving national second-city status and towards global competitiveness.

High immigration in the post-war period has meant that Birmingham is now one of the most ethnically diverse cities in the UK, with nearly 30 per cent of its population now non-white, mainly of Indian, Pakistani, and Black African origin. Major urbanization trends over the past 20 years have been population decline (especially due to the out-migration of affluent families), social polarization within the city's boundaries, the decentralization of homes and jobs and increased commuting, investment in roads and motorways rather than public transport, and radical changes in local government structure and organization (Murie *et al.*, 2003).

Urban discourse: economic restructuring, planning and the environment

As an industrial, developed-world city, Birmingham has followed the well documented trajectory of economic restructuring, urban re-imaging and city entrepreneurship. However, the role of the environment in such contemporary urban change has not been so widely analysed. The political and spatial economy approaches to competitive cities tend to underestimate environmental issues, leaving largely unexamined the bland policy commitments to sustainability characteristic of most urban restructuring strategies. Equally, environmental policy analysis has tended to take its own course, focusing on a different set of institutional actors through other theoretical lenses.

In this first section we analyse the ways in which the environment has been discursively constructed and given meaning in urban development policy in Birmingham, in particular in relation to economic restructuring. Planning policy is the main source of urban discourse examined. It is not, of course, the only relevant discourse, but does bear the authoritative weight of government and constitutes the locus of interaction with other sources of discourse. For the purpose of clarity of exposition analysis is divided into four periods, conveniences more than clear-cut divisions, each of which highlights the changing dimensions and significances attributed to the environment in accordance with the structural requirements of the

Location

Birmingham

West Midlands
metropolitan area

Central area
High-ethnic inner city
Mixed housing/social groups
Affluent suburbs
Main parks

5.1 Map of Birmingham.

advancement of economic restructuring in the city. We are especially interested in questions concerning the significance of business-led urban development, the role of the environment in the recreation of a sense of urban quality and the influence of national and international environmental issues on local policy.

The demise of planning and the eclipse of the environment (1980–85)

The first half of the 1980s witnessed a major assault on local government and the local planning system. The Conservative party victory in the general election of 1979, and the birth of the Thatcher era, occurred after several years of national economic decline, workplace unrest and frustration with irresolute Labour governments. The new Conservative government immediately set about a widely documented revolution in economic and social policy based on a neoliberal doctrine of a reified market. It was an ideological shift away from state intervention and towards private enterprise and competition, which demanded a frontal attack on all forms of social institutions and political opposition that might obstruct that change. Labour organizations and local authorities were amongst the priority targets. The latter, especially the metropolitan councils, were bastions of Labour opposition and local authorities in general were seen as out-dated bureaucratic machines and a major hindrance to the energies of private enterprise and individual initiative. The rhetorical and legislative attack on planning as a particular form of restrictive practice has been well documented (Allmendinger and Thomas, 1998; Atkinson and Moon, 1994; Thornley, 1991). Economic and social policy in general was enforced with political ruthlessness amid social conflict, and one of the many casualties was the idea of municipal-led, socially oriented, planned local development.

When Thatcher came to power, the Labour-controlled city council had already spent many years preparing a structure plan. Structure plans, a responsibility of county councils (Birmingham lay at the heart of the West Midlands Metropolitan County), were the pinnacle of the rationalist thrust of the British planning system. They were long-term, strategic spatial development proposals that presupposed relatively stable conditions, an interventionist sate and confidence in the ability of public institutions and rational analysis to direct the course of steady incremental growth. The 1980 structure plan, produced less than a year after the change of national government, meant the immediate and head-on collision of a bureaucratic and rationalist model of planning with a new free enterprise political culture.

The decline in regional economic fortunes in and around Birmingham and its social consequences were already the central concern of the structure plan. The depth of the economic crisis was considered sufficient reason for a 'substantial review' of the structure plan in the second annual review undertaken in 1982. It also brought to evidence the political difficulties facing the implementation of this pre-Thatcherite planning model. It focused in particular on the problem of finance resource availability for policy implementation and argued that 'the problems facing the West Midlands are of such depth that change in central Government policy and resource allocation are necessary', and the report is peppered with complaints about cut-backs and controls on local government spending (West Midlands Metropolitan County Council, 1983: 7). In retrospect, this was a hopelessly unrealistic dying call from the modernist tradition in British planning, soon to be forced into almost total capitulation by the abolition in 1986 of both metropolitan structure plans and the usually Labour-controlled councils responsible for them.

Nevertheless, the arguments behind that modernist appeal for a restoration of local authority power and money were substantial. The review describes the West Midlands county, with Birmingham at its visceral centre, as 'facing the worst economic crisis in its history, with growing social problems and tensions' (West Midlands Metropolitan County Council, 1983: 8). It describes how the traditional economic base of Birmingham in the engineering and metal-working sector was collapsing, unemployment was soaring, housing conditions were deteriorating, public transport was under severe strain, and dereliction and decay were blighting the urban landscape. This depressing urban condition, coupled with prolonged attacks on local government and the local planning system itself, produced a not unreasonable pessimism in the face of a vertiginous 'spiral of decline'.

This decline was reflected in intensified form in the physical environment. Changes to the planning system from 1980 onwards were designed to reduce delays in the development process and curtail the social and environmental scope of planning. Planning controls were reduced generally, taken out of local democratic control through the creation of Urban Development Corporations and minimized in the newly created Enterprise Zones and Simplified Planning Zones. The initiation of a 'property-led' development was part of a general trend of privatization, including public services, the massive sell-off of public sector housing and the 'commodification' of planning control (Tewdwr-Jones and Harris, 1998). The social function of planning was systematically prohibited through planning legislation and ministerial directives. All this was occurring in the midst of mass unemployment and job instability, regressive taxation and welfare reform, a long and violent confrontation with organized labour (especially the Miners Union) and a jingoistic war with Argentina over the Falkland Islands (or Malvinas). It is hardly surprising, then, that in this turbulent and divisive period of regressive social change, the environment was not high either on the political agenda or in terms of public concern.

Nevertheless, city planners in Birmingham were extremely concerned. The first few years of deregulated property-led development resulted in an accentuation of environmental deterioration without producing significant improvements in either the city's economic performance, urban regeneration or the quality of urban life offered to citizens. The 1983 structure plan review in Birmingham argued for the revision of the strategy founded on the regeneration and renewal of older inner city areas of economic decline. Contrary to central government policy, it proposed a greater emphasis on environmental and social objectives and, in particular, that 'environmental improvement is [therefore] the key to the longer-term regeneration of the older urban areas' (p. 11). This position was further substantiated by a critique of the effects of a narrowly economic approach to urban redevelopment in the city:

The over-riding emphasis on economic development in the inner city in the [central government] approved Plan has tended to be at the expense of the environment. The priority accorded to the needs of industry has not been conducive to improving either the attractiveness of the inner city areas, or the quality of life for people living or working there, nor has it assisted in the retention of population in these areas.

(Structure Plan Review, 1983, paragraph 3.29)

But what was the 'environment' in this context? The emphasis was quite clearly on the physical decline of the old industrial and socially deprived inner city areas, a persistent theme in post-war planning in the UK. The loosening of planning controls had achieved little in the way of reviving local economies and even less in terms of physical improvement. The early Conservative government considered not only the social redistribution aims of planning but also its meddling in design matters to be outside its proper remit. It was also notoriously adverse to activists of all kinds, particularly environmentalists, was nationalistic and had little time for the growing environmental bureaucracy within the European Community. Margaret Thatcher's personal environmental consciousness was known to be limited to an irritation over litter and her famous remark about there being no such thing as society could have equally applied to the environment. However, in the city of Birmingham different understandings were developing.

The environment as image (1986–92)

The intractable problems facing modernist city planning reached a critical point in the mid-1980s. In Birmingham, this was intensified by a change in political leadership over the period 1982–84, when the city council went under Conservative party control for the first and only time in the post-war period. This political change sharpened tensions between rival schools of thought within the planning department and saw the ascension of a new group more sympathetic to the free market precepts of the Thatcherite approach to local government and planning. Also, the very role of the city planning department was transformed. It could no longer hold claim to be the centre of strategic thinking, which was effectively transferred to the non-accountable private sector – consultants, think-tanks, quasi-governmental bodies – under the new ethos of privatisation and partnership sweeping across the political landscape. The change in political leadership was also important in the city, lobbying hard to achieve national Assisted Area status in 1984, a prerequisite to qualify for European Community Regional Development funds, as well as the opening of a city office in Brussels (Atkinson and Moon, 1994).

Birmingham was one of the earliest UK cities to adopt an entrepreneurial approach to urban development. The thinking behind such an approach was that cities had to actively attract private sector investment and create jobs in the new global economy. To achieve this 'boosterist' strategies were adopted in the form of publicly financed 'flagship' projects which would give the city a competitive edge. European Regional Development Funds as well as national and city budget allocations were vital in this respect. As a result, technical expertise of the traditional type became less important than strong political leadership, now aligned to the market knowledge of business leaders rather than property developers alone, a strategy based substantially on earlier experiences in the USA (Ward, 2002).

In the case of Birmingham, the first significant step in this direction was a public–private think-tank which produced the so-called 'Highbury Initiative' of 1988 (Birmingham City Council, 1989). An important outcome was the Birmingham Urban Design Study for the city centre, prepared by consultants Francis Tibbalds. The motive for this study, in an area of professional expertise then largely undeveloped within the local authority itself, was the need to address the continuing deterioration of the physical environment and its perceived effects on the attractiveness of Birmingham for inward investment. The city became aware that it had a disastrous image problem. Modernist planning, and modernist planners within the local authority, were seen as either directly responsible for or incompetent to deal with the new challenge. The urban design study laid the ground for a new approach to the city centre, based on a revival of traditional urban space (the street and square), intensity and mix of activities, the resuscitation of heritage, pedestrian comfort and a 'softer and greener' environment. The results were incorporated into a formal document on local authority policy for the city centre, the City Centre Strategy (Birmingham City Council, 1987). The environment was now firmly positioned at the centre of strategic thinking:

> The basis of planning in Birmingham and the City Centre in particular, is founded on the twin concepts of economic revitalisation and urban regeneration. The City Council believes that the quality of the environment and the economic health of the city have a direct effect upon each other, and wishes to see Birmingham develop as a better place in which to live, work, shop, or just visit. The Strategy seeks to attract activity, investment and ultimately development and job opportunities. The intention is, therefore, to make the City Centre more attractive, not just in terms of the physical environment, but also in a magnetic sense – to attract activity, investment, development and so on. However, the intention is not to attract activity for its own sake but to make the City Centre a better place – in the first instance, for the million people who already live within the City.
>
> (Birmingham City Council, 1987: 2)

The environment was no longer simply the object and finality of planning but now the condition for unleashing the dynamics of free-market development. The environment had to attract and seduce. The attraction of external agents was the essential ingredient of a marketing strategy in the emerging era of 'selling the city'. The planning strategy for achieving this had four main elements: better access, safe and stimulating physical surroundings, greater diversity of metropolitan activities, and the promotion of the seven different sectors or 'quarters' making up the widely defined central area, along with office and retail development to consolidate the city's role as a regional centre. Throughout, it was insisted, '[t]he quality of the environment is a key indicator of a successful City Centre'. For its part, seduction was aimed at the city's internal population. A revitalized and attractive city centre environment would not only edify a new collective image but also, it was argued, offer new opportunities for cultural consumption and urban social life.

By the end of the period early results around the first phase International Convention Centre quarter were materializing and Third City Centre Review (Birmingham City Council, 1992a) was replete with the statistics of economic achievement. However, the new environment-image proved to be both exclusive and costly. Whilst the city centre strategy made loud and frequent announcements to the effect that it was aimed at citizens and improving their quality of life, it is evident that the conception of 'quality of life' had a high dose of 1980s yuppie consumerism. The environment had been conceived and manipulated as pure image, in almost total disregard to the ecological considerations and social themes of sustainable development that had been gaining considerable momentum at national and international levels.

Internationalizing the image: sustainability mark 1 (1993–97)

The build up to the Rio Earth Summit of 1992 had stimulated greater environmental attention in the UK. Margaret Thatcher began to express nominal concern over global environmental problems and a national policy document, *This Common Inheritance*, was produced in 1990. Government ministers, politically shaken by the high Green Party vote in the 1989 European Parliament elections, also became concerned over citizen reaction to local environmental issues (Newman and Thornley, 1996). Weak environmental policy in the UK was based on market solutions and business responses, given academic respectability by London School of Economics professor David Pearce (1989), but there was a new internationalism in the air on environmental issues which central government could not ignore and the city of Birmingham was eager to exploit.

At the urban level, the 1991 Planning and Compensation Act required local authorities to include a section on environmental sustainability, and indeed many responded with much more enthusiasm than central government to the challenge of Local Agenda 21s as advocated at the Rio Summit. Birmingham produced its new style Unitary Development Plan in 1993, which largely reiterated the city's already established environmental foundations of economic revitalization and urban regeneration. Nevertheless, two important conceptual innovations were outlined in the 1993 plan around the notion of quality. Quality continued to be the 'keynote' to the city's environmental strategy, but the plan gave explicit recognition to the changing national and international economic landscape and the 'upsurge in awareness of environmental issues at both a global and local level' (p. 18). What is insinuated here, and almost completely ignored in the rest of the plan, is an acknowledgement of the ecological dimension to the environment and global ecological problems. The plan never mentions the word 'sustainability' but in the immediate aftermath of the Rio Summit it is clearly hinted that an attractive physical image will need the rational underpinning of ecology and sustainability discourse.

Some groundwork had already been done. For example, the city's first policy impact assessment on the environment had been undertaken in 1989 and in the early 1990s the city produced *Birmingham, A Greener Future*, a plan for natural habitat improvement on council-owned land. Additionally, the city's *Conservation Strategy* (Birmingham City Council, 1992b) had been revised and given greater priority. A year later the *Green Action Plan* was produced and an environmental forum set up. The *Green Action Plan* (Birmingham City Council, 1993) was a corporate programme to monitor the city's environmental performance, detail policy aims for key environmental issues (such as energy consumption, building efficiency, green purchasing, integrated pollution control, waste management, environmentally friendly transport, water quality, wildlife and flora, and environmental education). It also signalled the introduction of the new lexicon of environmentalist spatial management, such as eco-management, environmental audits, and the promotion of environmental awareness amongst business and citizens. These new topics and techniques were framed within the embrace of global environmental problems such as trans-frontier pollution, the greenhouse effect and climate change, and CFCs and ozone holes, supported by national and European policy documents and directives.

The environment as an international issue became important to city promotion. In 1993 the city, along with London and Manchester, was invited to respond to central government's City Pride initiative 'to define in a City Prospectus their visions for their city over the next ten years, and the actions each participant will take to achieve that vision'. The prospectus, entitled *Moving Forward Together* (Birmingham City Council, 1995) was published two years later as a glossy promotional brochure for an international market, ideally suited to communicate the city's new-found awareness and commitment to global environmental problems and sustainable development. The conception of the environment and its role in urban regeneration was substantially the same as in the 1993 UDP, although energy consumption (in transport, buildings, waste management) and public health issues have a higher profile. However, since one of the functions of this partnership plan was to attract European investment, a more 'worldly' environmental perspective was also required than that presented in previous plans, now legitimated in humanism and altruism. The Prospectus announced a set of six 'values' which underpin the city vision, all expressing a sense of common purpose: a 'one city' unity of stakeholders, equality of opportunity, sustainability, enrichment of the human spirit, cooperation through dialogue and, somewhat curiously, evaluation. This context of the introduction of sustainability is important not only in the sense of its presentation and associations with values, but also its more specific definition:

> Decisions taken now must consider not just current needs, but the needs of future generations. In developing a strategy to shape the city over the next ten years, we must be aware of the consequences over a much longer time scale. Whilst we all want to be part of a city which is improving both in terms of economic prosperity and social well-being, this short-term improvement cannot be at the cost of the city's long-term survival. This

is part of what the vision statement means by a 'thriving' city. We must aim for sustainable growth – and this means strategies which limit the consumption of non-renewable resources, which reduce damage to the city's natural environment and which create conditions that help maintain a *healthy, enthusiastic and committed* population.

(Birmingham City Council, 1995a: 9, our italics)

The prospectus was also part of a systematic attempt to introduce the environment into economic competitiveness and secure business involvement. In the Economic Development Strategy (Birmingham City Council, 1995b) sustainable economic development was one of a total of 13 strategic aims, under the issue of improving business performance. Here sustainable development was argued to be not only a city vision, but also a basic requirement of international legislation, corporate responsibility, commercial pressure and public opinion.

By 1998 it was widely acclaimed that the city of Birmingham had arrived on the international stage, and the environment may be seen as an important stepping stone to achieving that status. Already the venue of more international sporting, cultural and business events than any other city in the UK outside London, that year Birmingham hosted amongst others the Eurovision Song Contest, a meeting of the G8 group of nations and, coinciding with the latter, its own City Summit (Birmingham City Council, 1998). City leaders were invited from Chicago, Chongqing, Leipzig, Lyon, Johannesburg, Milan, Toronto, Barcelona, Frankfurt, St Petersburg and Yokohama, and the environment was one of the four 'big issues' of common international concern to be discussed. From an early internal preoccupation with spatial decline and decay, the environment was now part of the global discourse on sustainable development circulating amongst competitive international cities, business corporations and development institutions.

The second conceptual innovation established in the 1993 Unitary Development Plan was internal and social: that enhancing the environment was particularly important for improving the quality of life of 'the less well-off'. The Environmental Forum set up that year engaged industry, academics and NGOs, but not the community and certainly not the urban poor. Property-led development of the 1980s had produced new urban fragments and pockets of affluent physical environments without making significant impact on urban sustainability and social deprivation. Urban development was prone to business cycles and experiments with regeneration partnerships in the 1990s had little success, with reports by the Audit Commission and other research recommending greater integration of programmes and more involvement of local government, voluntary associations and the community (Newman and Thornley, 1996).

Sustainability mark 2: morality and performance (1998–2003)

Over recent years Birmingham's growing involvement in European urban programmes, city networks and funding arrangements enmeshed the city inextricably into sustainable urban development policy. The Istanbul City Summit of 1997 was significant in provoking this new agenda in Birmingham. In that same year Birmingham produced its own Local Agenda 21 called *Living Today with Tomorrow in Mind* (Birmingham City Council, 1997) and, a year later, its own set of sustainability indicators. It joined the Eurocities association of metropolitan cites and began to take part in a number of European Union supported initiatives such as Emas (eco-management and audit schemes), Pre-Sud (peer review of urban sustainability strategies), Integaire (implementation of the EU directive on urban air quality) and the European Common Indicators project on the quality of urban life. These international policy initiatives were incorporated into Birmingham's internal planning and development agenda through such things as the Sustainability Strategy and Action Plan 2000–2005, the creation of the Environmental and Sustainability Partnership and an environmental assessment and modification of the unitary development plan in 2002. The new major city centre regeneration project at Eastside aimed to 'set an example in terms of sustainable design, construction and long-term management' (Birmingham City Council, 2004a).

This now institutionalized internationalism signalled a decisive discursive shift in the construction of environmental meaning. On the one hand, whilst the relation between economic regeneration and environmental quality persisted, for example in the revised Unitary Development Plan, it was no longer the principal argument and as a result big business, property development and the environment of prestigious architecture became somewhat detached from the main thrust of the city's environmental policy. The economic sphere took an argumentative back seat as environmental discourse focused on two new strategies: one strongly related to the global ecological agenda and the other emerging from UK urban social policy.

Up until this period global ecological concerns had received only nominal attention. Local environmental NGOs complained of not being able to get the message across on the 'big issues' (Crean, 1998) and the lack of any real action produced growing frustration within the city's Environmental Forum. However, after 1998 there was a clear restatement of the importance of the environment. Now established in European city networks and with fast appreciating international credentials in the service sector, Birmingham adopted a sense of global responsibility in the framing of its environmental and development policy. It no longer argued in terms of self-interest but, rather, as a mature member of global urban society fully aware of the ethical, trade and development implications of its own economic survival strategy, as the following extract from the Sustainability Strategy and Action Plan 2000–2005 illustrates:

It is widely accepted that we are using up the world's resources at an alarming rate. If we are to survive and prosper into the future, we must take action now. We recognise that certain critical global issues such as population growth are beyond our immediate control. However, we must always remember that as one of the richest countries in the world, we are consuming a disproportionately larger quantity of the world's resources, which in turn means that we are causing larger adverse effects on the environment. Indeed, if we were to be self-sufficient within the city boundaries, Birmingham's land mass would only support 6,000 people. We recognise that everything we do has an impact and we recognise our obligations to ensure improved equality with developing nations.

(Birmingham City Council, 2000a: 4)

It would be easy to be cynical about Birmingham's abrupt moral turn to global sustainability, especially when in the same breath it also recognizes that through the European Sixth Action Plan Framework (coincidentally for the same 2000–2005 period), 'large quantities of European money have been put aside to fund sustainability actions'. It would also be grossly unfair, as many politicians and officials are undoubtedly genuinely convinced of the importance of ecological issues. However, the underlying explanation lies in the city's participation, through internationalism, in the demanding discursive community of multi-scalar institutions promoting global sustainability. In other words, it was a discursive obligation.

The second strategic change lies in the focus on community and the social agenda mentioned earlier. National government policy under New Labour became formally more committed to both sustainability and the regeneration of deprived urban areas (Rydin and Thornley, 2002). Nearly two decades of property-led development, privatization and local government restrictions had resulted in growing social polarization and racial tension in UK inner city areas. Deprived-area regeneration was aimed at addressing the problems of social exclusion and spatial segregation, and environmental discourse adjusted its conceptual structure and argumentative strategy accordingly. The city administration committed itself to putting its own house in order and argued for the sharing of responsibility and the duty of all citizens to act with 'courage, creativity and compassion' to meet the challenge of sustainability. '[W]hereas the environmental agenda was limited to what might be called the non-human world, sustainability is as concerned with employment, leisure, equity, health and freedom from crime as it is about protecting biological diversity and reducing air pollution' (Sustainability Strategy and Action Plan 2000–2005, Birmingham City Council, 2000a: 3).

This Local Agenda 21 expresses clearly the bifurcation of environmental policy in Birmingham: natural resource use and global ecological problems on the one hand, and local social issues and their management on the other, along with the displacement of the economic development rationale of earlier years by a moral discourse. However, this moral discourse implied practical obligations. Sustainability, with its emphasis on target setting and performance monitoring, enmeshed perfectly with New Labour's obsession with target setting and the monitoring of public service performance and 'value for money'. Sustainability required social as well as environmental indicators and incorporated the city's social programmes into the heart of sustainable development delivery. As such, sustainability became an extended form of measuring and monitoring both the city administration's performance and moral and behavioural standards of citizens now inextricably implicated in Birmingham's sustainable development policy.

Policy as discourse: constructions of the environment

A conventional analysis of environmental policy change in Birmingham would focus on the city's narrow early experiments, the obstacles constituted by UK national policy limitations, the importance of a European framework, the gradual incorporation of ecological and social dimensions of sustainable development, increasing technical expertise in environmental management, and so forth, as if tracing a linear path towards rational ecological enlightenment and responsibility. However, in examining policy as discourse we have attempted to draw out an alternative logic and a complementary account based on the discursive construction and reconstruction of the environment. We argue that the environment was not something constant and external to be discovered and then rationally managed. It had to be discursively edified and given meaning, a process which unfolded within the dynamics of the overall city problematic and, especially, the way in which Birmingham responded to the challenge of economic restructuring and globalization.

In the four schematized periods examined above we outlined the main characteristics of this construction and reconstruction of what the environment was, its constituent components, its meanings for city development and significance for the different actors and social groups. Superficially, it suggests an evolutionary process and a gradual shifting of attention from the city centre physical environment to the wider social environment, from artefacts to natural resource systems, from local to global development issues, from economic to moral and behavioural concerns. However, we argue that this was dependent not on lineally progressive knowledge of the environment, but rather on non-environmental discourses, external events and different conjunctures in the city's development.

This, in turn, depended on the imprecision of the notion 'environment' which embraced three discrete sets of objects. First, the environment in Birmingham denoted a set of physical objects such as buildings, urban space and infrastructure, or the artefacts of the city. Second, it included a set of local facilities such as schools, shops and recreational facilities, or a set of social amenities. Third, it incorporated the elements of natural resource systems such as air, water, soil, and natural habitats, or a set of ecological components. This reflects the British urban cultural and planning tradition. On the other hand, the stable yet vague notion of what the environment *is* contrasts sharply with the effervescence concerning what it *means* and why it is important and for whom. This was dependent upon the dimensions or sets of objects of the environment being assembled at an abstract level through terms such as the 'general environment', 'atmosphere', 'ambience' or 'habitat' whilst simultaneously being chopped up into the 'working environment', 'living environment', 'business environment', and so forth. The environment then became a metaphor for differential spatial consumption according to the interests of particular social groups – groups of business people, groups of workers, local residence groups, groups of investors and tourists, and so forth. Policy discourse then fulfilled the task of establishing meanings which assign privileged rights of access to and control over portions of the environment.

In Birmingham, the 'prime site' of the environment has been the city centre, discursively assigned to the private sector. This 'business environment' has been consistently presented as one of life and death, not in any ecological sense but in terms of economic survival in the new global order, vital to investment attraction and the restructuring of the city's economy. This discursive privatization of the environment focused on the city centre but also extended into some adjoining inner city sites and green belt areas. It established development opportunities as environmental responsibilities which only the private sector could assume with the necessary knowledge and resources. This did not prohibit public funds being poured into promoting private profit-seeking through urban redevelopment in the name of the city business environment. But how important is the environment for business? Experience in Birmingham tends to confirm the secondary status of the environment in relation to other attraction factors such as financial incentives, development costs, labour market conditions and infrastructure provision, especially for manufacturing industry. This is not to say that the environment lacks economic importance altogether, but in development based on the service sector, business tourism and the cultural economy, the importance of the environment acquires a very particular significance: the environment as image, its aestheticization. The following opinion of an executive of Birmingham's National Exhibition Centre is quite clear in this respect:

> As you know Birmingham has got a very high standard of environment, it's a clean city. I guess visitors' concerns would be more about personal security, whether the safety of visitors is assured, the perception of that, and from our research in America we think we have achieved that reputation [. . .] The quality of the environment is very, very important for the marketing of events. Conferences, big exhibitions, sporting events, staging the Eurovision Song Contest, whatever you think of, the end result has to be one of a very good quality impression, good television, good photographs, good environment for the delegates or the participants or the spectators, and it's a very tough one to get right. You don't want to build palaces for the sake of it, but you do need a good environment and ambience created, you can't just build a shed or cheap box and somehow expect the market place to become excited.
>
> (personal interview, 1998)

In its radicalized form, the business environment is transformed into a question of promotional image and spectacle, where the ultimate measure of achievement is the market rather than social criteria or civic judgement. Certainly, there is only limited room in dominant 'business environment' discourse for ecological considerations or for those sectors of the city and urban populations which fail to take part in the new cultural economy. In the following section we examine how the discursive alliance of strategic planning partners established itself across urban policy in Birmingham and constructed the environment as a significant issue for the rest of the city.

Institutions: the regulation of environmental quality and welfare

In the previous section we examined the change in general meanings ascribed to the environment in urban policy discourse in Birmingham, especially through the different kinds of development plan, and in passing indicated the institutions involved in endowing such discourse with strategic authority. However, as we have argued earlier, institutions are not merely technical bodies for detailing policy and implementing change; they simultaneously perform the function of providing substance to general meanings, authorize who can participate and how, and materialize general discourse in terms of concrete action. This section will focus on the institutional mechanisms by which general environmental meanings are introduced into urban processes. Our understanding of institutions includes all those formally recognized organizations of government, business and civil society which participate in the urban policy process.

Before beginning a more detailed analysis, it is useful to review at a theoretical level some of the features of institutional change identified in the preceding description of urban policy in Birmingham. It is widely accepted that neoliberalism is not simply a withdrawal or retrenchment of the state and the deregulation of the economy, but rather a question of system transformation requiring new strategies of state intervention. In the transition from post-war welfare-statism to neoliberalism, Jessop (2002) identifies four principal characteristics of the 'régime shift' in North American and European contexts:

1   the promotion of international competitiveness and sociotechnical innovation,
2   the debilitation of the national scale of policy-making,
3   the reliance on partnerships, networks, consultation, negotiation and other forms of reflexive self-organization as opposed to top–down or bureaucratic systems of interaction, and
4   the subordination of social policy to economic policy, labour market flexibilization and the understanding of the social wage as a cost of production rather than a means of redistribution of wealth and social cohesion.

Such changes have been characterized in terms of an initial 'roll-back' phase of deregulation and privatization in the 1980s, followed by a 'roll-out' phase of re-regulation and institutional reconfiguration from the 1990s onwards, with cities in particular becoming 'geographical targets and institutional laboratories for a variety of neoliberal policy experiments' (Brenner and Theodore, 2002). Far from being a stable or even coherent process, system change requires continual experimentation in regulatory fixes concerned not only with the apparatus of production but also the construction of notions of citizenship and the extension of new régimes of accumulation into everyday urban life (Keil, 2002).

The analysis of urban development policy discourse was concerned principally with the first of Jessop's four régime features. However, the adequate understanding of urban environmentalism needs to take into account the whole range of 'regulatory fixes' across urban and social policy, and institutions provide a convenient way of penetrating a complex field of analysis. Our interest is to highlight the way in which transformations of the institutional regulation of urban development created a quite specific place and role for the environment in Birmingham, for example its significance for the city's economic restructuring and, later, social well-being.

The state, planning and the environment

All radical neoliberal projects involve a centralization of power and authoritarian national government in order to impose unpopular measures and overcome local resistance. However, apart from the abolition of the metropolitan county councils in 1985, the Conservative government of the 1980s made no significant changes to local government organization. Rather, it modified the mode of operation of the relations between central and local government through stringent controls over public finance. This was an extremely effective method of not just reducing overall public expenditure but also exerting detailed control over how reduced local budgets were used. Given that local initiatives were then restricted to central government priority areas, local authorities were faced with either succumbing to central government directives or entering into a direct and costly political confrontation.

Modifications made to the planning system provide a good illustration of the way in which central government controls were harnessed to a deregulation strategy at the urban level. Thatcherite neoliberalism asserted the unmitigated supremacy of the market as a means of creating and distributing wealth, whilst the extant planning system was a product of the post-war welfare state and its explicit commitment to the control of the means of production and social redistribution. For Thatcherism, planning was not only held to be constraining the economy but also contributing to social decadence and a 'dependency culture'. As a consequence, one of the first measures of the Conservative government was to extirpate planning of its traditional social aims. In the early 1980s it was made clear that plans should restrict their concern to land use matters and avoid positive discrimination, for example for low-income housing or through reference to grants and assistance programmes. Failure to comply with this advice would be remedied through central government administration of the appeals procedure; for its part, design was a matter for developer and customer (Circular 22/80). Municipal plans would have priority over strategic regional ones as part of a general speeding up of both plan-making and development control procedures. Economic growth and the interests and knowledge of developers were to have precedence over the public interest as the main criterion or 'material consideration' in planning decisions, and so forth (Thornley, 1991). The deregulatory shift in favour of developers was accompanied by the huge rise of 'quangos' (quasi non-governmental organizations), appointed and controlled by central government for the management of a wide range of institutions across the social services in areas such as education, health and housing.

In this political take-over of local institutions by business and central government, the environment was systematically eroded as a legitimate concern of municipal planning and institutionalized local politics. On the one hand, the drive to speed up the planning process and reduce delays in development control decision-making implied the minimization of public consultation and the restriction of officer discretion. On the other hand, design issues were held to be beyond planning's administrative remit and professional competence. Thus, both opportunity for the expression of public concern over the environment (through participation) and control over the construction of the environment (through design control and advice) were significantly restricted. The only exception to this rule concerned explicitly denominated conservation areas relating to both the urban and natural environments. While this deregulation was going on with regard to development and the environment, re-regulation was occurring in other policy fields. Drastic restrictions were imposed on trade union activity and the right to gather in public, law and order was made more punitive, rhetorical attacks were launched on lone-parents, drug-users, Romany and New Age travellers and other social 'deviants', and tighter access to and conditioning of the social security benefits system was enforced.

National policy and public opinion did address the environment and sustainability more systematically in the 1990s. However, the tendency of post-Thatcher administrations to move towards less confrontational government and a partnership approach was arguably more significant in opening up institutional space for the environment than any explicit national policy on the environment as such or subsequent minor reforms to state territorial organization. Even during the 1980s it has been demonstrated that statutory reforms to the planning system were consistently circumvented by local government, and this seems to be the case in Birmingham. City politicians and planners became acutely aware of the environmental deficiencies of deregulated urban development. Deregulation was neither bringing significant economic gains for the urban economy nor improvements to the physical environment. We have seen how, in the mid-1980s, the city authorities in Birmingham insisted on the environment and design as the key to economic recovery through city centre renewal. The re-imaging of the city which began in the mid-1980s with the International Conference Centre was subsequently extended to the whole of the city centre, to the extent that the notion of environmental quality became the key word in urban development, for which the city itself assumed strategic responsibility.

Birmingham, like many other cities in the UK, was way ahead of central government policy in this respect. Only in 1994 did the Department of Environment produce its discussion document *Quality in Town and Country* followed by the Urban Design Campaign of 1995–96 oriented towards design-led urban regeneration. The generally small-scale projects developed under this central government initiative, one of which was carried out in Birmingham, were referenced specifically to sustainability, although there was 'little sign that more compact and sustainable urban living is being either promoted or achieved' (Chapman and Larkham, 1999).

In the post-1997 New Labour governments, two modifications of significance were introduced to local government: the option for local authorities to adopt a cabinet system of government headed by a city mayor (as opposed to the traditional council leader and committee system) and the extension of the partnership principal into all areas of urban policy implementation. The first, proposed but not fully adopted by Birmingham in 2002, allowed the city to formalize its tradition of strong political leadership and develop an executive-style city administration. The second obliged the city to focus its attention on the decaying areas of the working class and ethnic minorities through urban regeneration programmes. Furthermore, the regional level of policy was bolstered through strengthened, though non-elected, regional offices for economic and spatial development, including environmental regulation. In the process, the environment was further consolidated as a question of 'urban quality' and 'quality of life' within the strategic aims of economic restructuring and urban competitiveness. However, this latest step in the repositioning of the environment in urban development was part of general rescaling of urban regulation, as the implications of supra-national levels of government also began to exert their full influence.

Scales of government and urban environmental policy

Earlier sections have described how urban policy in Birmingham became increasingly international in outlook. From the mid-1980s onwards the then European Community was seen as an important source of funding to help rebuild the city's economy, which in turn inspired the city's wider presence in international organizational networks and an early response to the requirements of global competition. In the early 1990s something similar occurred with regard to the environment, as European environmental regulations and urban policy began to influence the way the city presented itself in the international arena. In both the economic and environmental fields, development initiatives in the city were substantially contextualized by international rather than national policy.

Theoretical explanations of this sort of change have emphasized the importance of globalization in terms of the re-scaling of urban policy. Globalization brought with it the weakening of the nation state as the primary spatial scale of economic regulation, to be replaced by global institutions such as the World Trade Organization and the International Monetary Fund and regional trading blocks such as the European Union. As global economic change focused on cities and city-regions, they became the locus of complex multi-scalar intersections of local, regional, national and international government (Swyngedouw, 2000; Brenner and Theodore, 2002). The success of urban entrepreneurialism was thus dependent not only on the character of a city's own internal initiatives and experiments with urban governance but also they way in which these interplayed with national policy changes and institutional reorganization, international policy networks and global regulatory mechanisms. It is generally agreed that the outcomes are not uniform across countries and cities but 'path-dependent' on inherited institutional and social landscapes. How did the rescaling of government affect the institutional regulation of the environment in the case of Birmingham?

One of the most salient features after the mid-1990s was the re-emergence of the regional level of government. City-regional authorities such as the West

Midlands Metropolitan County had been abolished in the mid-1980s, precisely when European development policy was centring on the regional level. City entrepreneurialism therefore was disadvantaged to the extent that access to European funds was made more difficult, although the case of Birmingham illustrates that such difficulties could be overcome through energetic lobbying. There was also a noticeable 'greening' of European Structural Funds for regional development in the Fifth Environmental Action Programme 1993–2000, with access to funds being conditioned by environmental criteria and appraisal. Also, in the 1990s central government became increasingly aware of the obstacles posed by a lack of regional government for the effective coordination of national sectoral policies, especially with regard to urban regeneration.

In the case of Birmingham and the West Midlands region, the Government Office for the West Midlands was set up in 1994 to coordinate the work of national departments at a regional level, and assumed regional responsibility for sustainable development policy. The West Midlands Round Table for Sustainable Development was created by local authorities in 1996 as a consultative body, along with the regional office of the national Environment Agency. Other ad hoc bodies and working groups were also established, such as Environmental Technical Group and later the Sustainable Development Technical Group in support of the West Midlands Enterprise Board (1997). In 1999 the Regional Development Agency, Advantage West Midlands, was established to promote economic development along with a Regional Chamber. In the absence of any directly elected regional government, these organizations are administered by a variable mix of central government appointees, nominated councillors from local authorities and representatives from the business community and NGOs, and each body is required to cross-reference its policy to that of the other organizations. This complex pattern of 'rational bureaucracy and distended democracy, together with the associated working groups, regional fora, research consultancy inputs, monitoring organizations and other locally based non-departmental public sector organization has come to be called "the regional spaghetti" by some officers (Charlton, pers. comm.) and "the regional omelette" (Crean, pers. comm.) by some environmentalists' (Pratt, 2000).

Nevertheless, some important general characteristics can be drawn. First, national policy on sustainable development (*A Better Quality of Life: a Sustainable Development Strategy for the UK*, DETR, 1999) ensures that regional environmental policy is subordinated to economic development. This is realized through the composition, responsibilities and relations established between the different regional agencies, notwithstanding the participation of more environmentally conscious actors, such as some local authority officers and NGOs representatives. Second, the environmental component of regional strategy tends to be circumscribed by the requirements of European legislation and funding rather than by any authentic regional sustainable development policy. Third, the complex intermeshing of formal policy statements establishes a strong 'policy régime' and 'discourse alliance', made still more impenetrable by the complexity of its institutional structure. Fourth, regional environmental policy lacks any firm base in representative democracy, as is the case with regional government as a whole in England.

Strategic regional responsibility for the environment brings with it the depoliticization of the environmental issues. This is achieved not only through the fact that regional institutions are non-elected and unrepresentative, important as this is. The institutional coding of regional environmental issues through reference to national and European policy directives involves a sophisticated formalization of environmental policy debate that excludes those uninitiated in the complex international bureaucracy of the environment. Not only the ordinary citizen but also non-specialist officers, academics and activists face formidable discursive barriers. The result is an effective institutional enclosure of environmental policy debate.

Furthermore, the regional definition of sustainable development policy not only locates the environment within and as a subsidiary to the economic development strategy, but also determines priority environmental policy areas, reinforced by the regional level coordination and control (by the West Midlands Enterprise Board and later Advantage West Midlands) of bids for UK and European regional development funds. In this way the contradictions of sustainable development are effectively removed from city politics. In turn, environmental management as undertaken by the city itself can focus on more practical concerns such as project development, service delivery and citizen behaviour: a depoliticized technical agenda, the legitimation of which depends on performance criteria. A good example of this is Birmingham's participation in the recent European initiative of Common Indicators (European Commission, 2004), designed to 'help towns and cities monitor their environmental sustainability' without putting underlying urban policy to any serious examination or political test. Furthermore, performance is increasingly a matter for the service delivery agencies, communities and individual citizens involved in specific area programmes, rather than a general municipal responsibility.

Competition, partnership and privatization of the environment

In the analysis of urban development policy in Birmingham we saw how the environment was discursively divided into segments (the business environment, the living environment, and so on), each of which were then 'allocated' to particular sets of actors in the urban development process. In turn, this process was regulated by new institutions resulting from multi-scalar policy intersections, as described above. Especially noteworthy was the emergence of a rational-bureaucratic style of environmental policy and regulation at the regional level, whilst at the city level practical performance became the main concern. Here competition, partnership and the culture of enterprise were the driving forces, leading to some quite specific institutional forms of appropriation of the urban environment.

Competition was, of course, the pillar of the Thatcherite neoliberalism, with the market principle preeminent not only in the sphere of economic production but also actively extending into public administration through such measures as compulsory competitive tendering, outsourcing and auditing. As far as urban development is concerned, the crude approach of the 'roll-back' 1980s was to rely on competition between private firms through deregulation and limited public-private partnerships, but a more sophisticated version evolved in the 1990s. To a large extent this was established through the employment of competition as the strategy for the allocation of public investment funds. The logic of competition was behind initiatives such as City Challenge (set up in 1991), the Private Finance Initiative (1992), City Pride (1993), the Single Regeneration Budget (1994), English Partnerships (1994) and the Millennium Commission's allocation of Lottery Funds (1995), echoed at the European level by the Regional Development Funds (1993). All of these required individual cities to compete amongst each other for public funds on the basis of especially prepared proposals. The competition principle was further stimulated by bidding for prestige international sports and cultural events, the European Cultural Capital awards and the growing use of national and international league tables on city performance, business competitiveness and quality of life. Institutionalized competition became the defining feature of the regulatory fix of the 1990s and after, with the specific requirement of public–private partnership (Jones and Ward, 2002). Rather than addressing city-wide needs, competition focused on the formation of coalitions of partners to develop area projects and mobilize investment in specific locations.

Competition and development coalitions appropriated Birmingham's city centre, which had been discursively established as the basis of economic recovery and principal location of the 'business environment'. A private consultant was employed to establish the general approach to the city centre's spatial development, and under the Highbury Initiative of 1987 the city centre strategy was formally adopted by an elite group of political leaders, local businessmen, the real estate sector and international experts, pushed along initially by the leader of the Labour controlled council. The first stage – the Broad Street corridor, a business and cultural sector based on the International Conference Centre, Symphony Hall and National Indoor Arena – relied heavily on public sector funding and strong local political leadership. Development was also facilitated by the mixed capital NEC company which had been established in 1970 to develop the National Exhibition Centre, opened in an out-of-town location in 1976. The NEC Group was set up to move the International Conference Centre development in the 1980s, and public sector expertise in this kind of development was important in creating private sector confidence and effective working relationships between city leaders, high-ranking officials and the business sector (Barber, 2001). Development of the ICC complex was undertaken with UK and European funds but also involved large-scale investment from the city budget, although later developments became less dependent on public sector financing from the city budget and more on specific partnership configurations for particular areas and projects.

The spatial framework for partnership projects had already been established through the 'quartering' of the city centre into business and cultural districts. For example, the Heartlands district in the north-east of the centre was conceived as a new business district and public commitment was limited to land assembly, but when private sector investment failed to materialize central government stepped in and Heartlands was declared an Urban Development Corporation. In the Eastlands quarter, conceived as an educational and cultural quarter, public investment in land assembly and infrastructure was dependent on national and international development funds, including the Millennium Point science and culture building financed through National Lottery Fund awards. Private investment and diverse development companies were then responsible for site-specific projects such as the Custard Factory for the creative industry sector (led by a London-based property company which had participated in the original Highbury Initiative). This was the general rule after the mid-1990s, including important developments such as the Brindley Place office district, the Mail Box upmarket designer shopping mall and the redevelopment of the Bull Ring shopping complex.

These partnerships and property-led development coalitions for specific quarters of the city were concerned with rebuilding the basis for a new city and regional economy aimed at the financial, business tourism, retail and culture industries. However, this extended partnership approach did not lead necessarily to a restoration of local democratic and politically accountable decision-making. Partnerships with the private sector in urban development were extended formally to include civil society (academic sector and 'voluntary organizations') and closer working with the community (Highbury 3 Conference, 2001), although in effect they tended to result in the consolidation of new urban management elites formed by executive officers, leading politicians and the property and business communities. They increasingly extend their influence from the ownership and construction of the city centre environment, to control over its use, surveillance, maintenance and promotion. The environmental agenda was strictly controlled in this sense and the 'democratic deficit' of partnerships on large-scale development projects (Swyngedouw et al., 2002) also involved an 'environmental deficit', in the sense that the pursuit of profit excluded any integral, holistic sense of urban sustainability.

Whilst capital investment was being concentrated in spectacular fashion in the centre of Birmingham, the rest of the city, particularly the problematic 'living environment' of the poorer areas, was subject to a different regeneration dynamic. As in all big cities, social inequality and ethnic divisions in Birmingham had grown in the 1980s, and when urban regeneration did occur in the poorer residential areas it was not property-led but state-interventionist in nature. Local authorities, community organizations and the 'voluntary sector' were the main partners in trying to deal, in an area-based way, with problems of poverty, unemployment, drugs, crime, poor housing, declining public health and the lack of education and training opportunities. Inclusion and empowerment were the buzz words, as deprived areas were encouraged to organize as a community and take part in the complex process of competitive bidding for regeneration funds (Roberts and Sykes, 2000).

Work, welfare and the environment of the poor

Policy discourse in Birmingham constructed the environment as an issue of quality: initially as a quality requirement for economic restructuring, later as a key determinant of the quality of life of all citizens. We have already argued that this was not simply a felicitous conjoining of two separate discourses (economic and environmental) but rather an interdependent construction of meaning which legitimated a policy of differential spatial appropriation and the privatization of the city centre. This section explores the place-specific politics of the deprived sectors of the city or the 'living environment' of what became known as the 'underclass'.

In a general sense, place specificity has been an integral part of the reconstruction of the notion of welfare under neoliberalism. The old post-war welfare state was based on the principle of universality, under which space was little more than a distributional framework for meeting social needs. National institutions provided a uniform coverage, albeit with minor spatial variations. Over the past two decades, however, there has been a redistribution of welfare responsibilities between the state, the private sector (for example, health provision and recreation), individuals (for example, in housing and provision for old age), and the community (for example, the management of schools and hospitals, care in the community). This redistribution and spatialization of welfare highlighted the importance of localities. In turn, the profusion of area-based urban regeneration programmes in the 1990s provided the policy and institutional mechanisms required to consolidate the idea of welfare as a local concern and an integral dimension of the 'living environment'.

From the 1990s onwards spatialized welfare in the UK has been delivered through urban regeneration policy. This general term refers to a complex array of programmes, institutions and funding provisions based around the notion of local government working in partnership with other public agencies, business, the 'voluntary sector' and the 'community'. In deprived areas, these programmes focus on a mixture of employment, housing, health, education, environment, crime and safety issues, with the presumption that local solutions involving the mobilization of the community are more socially effective. The official aim of regeneration is to make welfare expenditure more cost-efficient, empower citizens through participation, articulate policy through 'joined-up' government, and extend the principle of partnerships into the field of social programmes, involving local organizations and the community. Introduced under Conservative governments, the most significant changes introduced under New Labour from 1997 were the strengthening of the aims of social inclusion, stake-holding and value for money.

By the early part of the new millennium there were 24 area regeneration programmes operating in the city (Birmingham City Council, 2004b), a novel factor over earlier urban policy experiments being the concerted effort to get people back into the labour market. Competition is again the key concept, as communities are required to participate in competitive bidding for regeneration funds and individuals are encouraged or coerced into becoming competitive in terms of skills and employability. It is difficult to assess their overall physical and social impact in Birmingham, but it is unlikely to differ radically from the national situation where, despite some notable results, criticism has focused on the slow, time-consuming process and the bureaucratic top–down approach. Regeneration programmes have been particularly ineffective in reaching black and ethnic minority groups; only four of the 555 bids approved nationally between 1994–96 came from such areas, including one from the Birmingham-based West Midlands Black Regeneration Network (Beazley and Loftman, 2001).

From our perspective, more important than the huge operational complexity and debatable concrete results of regeneration are its underlying spatial politics. Of particular interest is the instrumental function of regeneration in realizing the redefinition of and relations between work, welfare and citizenship. It has been observed that a major aim of urban regeneration programmes is getting people back to work, yet the social nature of work has changed radically under neoliberal social policy. At the discursive level, human needs and universal rights were replaced by the moral principles of self-reliance, personal responsibility and duty to oneself and to society as a whole (Powell, 1999). The low-paid and out-of-work, once victims, were transformed into villains whose access to welfare services became conditional upon a process of rehabilitation demanded by a more coercive welfare régime. Privatization and means-testing introduced in the 1980s re-initiated a more punitive system of benefits for those (especially the young, lone parents, the disabled and the long-term unemployed) who refuse to take part, for example, in New Labour's New Deal Welfare to Work schemes, and similar sanctions for the non-compliant have been extended to areas such as social housing and 'anti-social' and criminal behaviour (Dwyer, 2000). Employment incentives were also introduced such as tax credits, benefit enhancement for job-seekers and the introduction of the minimum wage. The overall effect was to construct work as the primary source of social inclusion, individual fulfilment and welfare entitlements (Hewitt, 2001), and area-based regeneration was a major implementation strategy and regulatory mechanism.

Urban regeneration programmes in the poor areas of the city were also part of a concerted effort to construct a new citizen in accordance with the demands of economic restructuring and urban competitiveness. Partnership, in contrast to participation, presupposes a different political subject, constituted through the exercise of juridically determined rights and obligations rather than political identity and representation. This new political subject was systematically constructed throughout the 1990s through such things as the Citizens Charter, a contract culture in local government, the permanent auditing of public service targets and delivery performance, and the definition of citizens as clients and consumers of privatized and out-sourced public services or of the remnants of state provision now run along private management lines with an emphasis on 'value for money' (Blackman, 1995). Urban regeneration was a mechanism for targeting marginalized communities, transferring responsibility to the poor and demanding the mobilization of their own resources of time and organizational effort in order to 'empower' themselves – into the lower end of the labour market.

In the wider urban context, urban regeneration needs to be seen as part of the rebuilding of the city's economy around the service sector, which required the construction of a new worker subjectivity, new incentives to retrain and harsher sanctions for the unemployed, in order to bring the city's population into line with the requirements of its new economy and image. In short, the new entrepreneurial city had to create new enterprising citizens, or at least citizens malleable to the requirements of the entrepreneurial ethos pervading the city.

At the same time, the city itself – as a physical entity and experiential framework for everyday life – had to contribute a sense of well-being to counterbalance the social and individual costs caused by the effort and disruption of economic restructuring. In this sense, the environment was both representation and substance. Capital-intensive city centre regeneration created a new 'imaginary' of the possibilities of urban transformation, whose procedural characteristics of partnership and competition were replicated in community-intensive form in the poor sectors of the city. Quality in the urban environment and the aesthetics of regeneration, materialized in the city centre, were extended to the working population through the idea of something to be striven for in community regeneration schemes. The citizen, newly disciplined into the requirements of flexible work régimes, accustomed to growing wage inequalities and job instability, was seduced into at least partial compliance by environmental aesthetic consumption. Hard work and effort (both inside and outside the labour market) could seek its reward in the pleasures of the physical environment; identity as worker its recompense in the new image and status of the city. In an individualized urban society in which welfare was a question of personal performance and recourse to state welfare rights a stigma, the environment was constructed as a field in which just rewards could be legitimately demanded and enjoyed, even by the most marginalized and disadvantaged citizens.

Institutions and the mobilization of environmental meaning

Before ending this section on institutions we shall attempt a brief synthesis of the institutional mobilization of environmental meanings in Birmingham and mention some possible theoretical interpretations. Certainly one of the most remarkable things concerns the way institutions fragmented the 'environment'. What are supposedly the collective and integrating characteristics of the environment as a general notion were dismembered by and in concrete institutional practices. Institutional configurations and the power relations they coalesce assigned and appropriated sectors of the environment on behalf of different politico-economic interests (especially around the spatial extremes of the property and labour markets) whilst at the same time constructing and adapting the environment's general significance in terms of the qualitative experience of urban space. Spatial consumption and its lifestyle implications, whether in the glitzy city centre or the deprived residential areas, became the main thematic argument. This aestheticization of the environment implied the relegation of ecological concerns relating to global sustainability (such as city energy and resource consumption, pollution, transport and waste, natural resource systems and so on) to the secondary plane of a technical agenda and the city's political backwaters.

Urban planning discourse had, of course, previously established a functional division of the environment in an abstract fashion ('working environment', 'business environment' etc.). However, what at first sight appears as a harmless piece of loose conceptualizing of space did in fact discursively open the way for and legitimate the appropriation and privatization of the environment through different sets of institutional arrangements. Thus, the partnerships established with the property sector in city centre renewal achieved a monopoly on the 'business environment' and the mobilization of environmental meaning through architecture and urban design. On the other hand, the mobilization of environmental meaning in the poorer sectors of the city was undertaken through urban regeneration programmes focused on the deprived communities and involved partnerships with other state and 'voluntary sector' agencies.

The high degree of institutional innovation and complexity at all spatial levels – regional, city and local – is a convincing illustration of the demand for experimental policy and regulation fixes under neoliberal urbanization, with the rescaling of government also influential in the creation of new institutionalized forms of economic/environmental policy setting and implementation. In this sense, institutions in Birmingham were clearly crucial in materializing the idea of sustainability as ecological modernization, whereby big (property) business was held to be both dependent on and the guarantor of a quality environment: the ostensible win–win situation at the urban level. However, the politics of quality varied enormously. Whereas the city centre environment was outward-looking and international, in residential areas it became a closed and coercive universe of welfare conditioning, social control and limited aspirations. The 'living environment' of the poor and ethnic minority groups became institutionalized around duties and obligations, the 'minor practices of citizen formation [are] linked to a politics of the minor, of cramped spaces, of action on the here and now, of attempts to reshape what is possible in specific spaces of immediate action . . .' (Osborne and Rose, 1999: 752). This interpretation of environmentalism for the poor throws a new and complementary perspective on the frequent criticisms of city centre environmentalism as being socially exclusive and encouraging gentrification. The social equity and ecological issues of sustainable cities have been latent and implicit up to now. We will explore them further in the following section.

Socio-spatial form: city greening up- and down-market

Sustainability is now high on Birmingham's policy and promotional agenda. Environmental issues, linked almost exclusively to economic restructuring and the city centre in the 1980s, now extend into all areas of city policy and spatial planning, supported and mobilized by a complex régime of institutions that insert the environment into every aspect of public life. This formal environmentalization of Birmingham has, of course, been undertaken as part of an extensive and radical transformation not only of the local economy but also of the city's image and architecture, forms of governance, urban social relations and their regulation, consumption patterns and city lifestyles, and so forth. The environment has played a key role in this transformation. Councillor Stewart Stacey, Chair of the Environment Committee argues that '. . . for the last 20 years the council has been involved in a massive sustainability project, that of re-cycling a whole city' (Birmingham City Council, 2002). In this section we will examine what sort of urban recycling has taken place and the nature of the city that is emerging.

In general terms, the way Birmingham has addressed environmental issues and more recent sustainability policy follows the European model of ecological modernization at the urban level. Environmental protection has been pursued along the lines of economic growth in the service sector, policy coalitions, public-private partnerships and management techniques which maintain the appearance of a win–win situation, whereby the rational incorporation of the environment into economic and social development leads to new forms of integral growth patterns. We will now look into the real city behind the policy rhetoric.

Recycling the city centre

The initial and still most important aspect of urban change in Birmingham has occurred in the city centre. Huge amounts of public and private investment have been poured into transforming the physical and aesthetic environment, erasing as far as possible the rapidly obsolescent architecture and urban form of 1950s and 1960s modernism and reinvigorating the traditional business and cultural economy. City centre redevelopment in Birmingham has been acclaimed as one of the more outstanding examples of urban entrepreneurialism and city re-imaging, although public perceptions continue to lag behind some of the concrete achievements.

It should be recalled that Birmingham had been at the forefront of urban and architectural modernism in the post-war period. The city centre in particular had been a vast exercise in redevelopment that, along with war damage, swept aside much of its Victorian past amid an adventurous and soon notorious ensemble of shopping complexes, tower blocks, inner city expressways and multi-storey car parks. The aesthetics and sociology of the 'concrete jungle' had been widely criticized and the Birmingham Bull Ring Centre and Rotunda building acquired national notoriety as symbols of the broken and now obsolete dream of urban modernism. Birmingham's dour image was not helped by its somewhat dull and provincial cultural life. In the 1980s, the city was lagging behind other major cities in the United Kingdom, such as Manchester and Glasgow, in the early stages of the race to establish a new and internationally competitive image of metropolitan, and more specifically European, sophistication.

The first phase of city centre redevelopment took place on the Broad Street corridor between 1987–1992, involving the International Conference Centre, Symphony Hall, National Indoor Arena and Hyatt Hotel, to which was later added with the Brindley Place office development. In architectural terms, this redevelopment of a previously run-down, mainly industrial area certainly provided the city with modern facilities aimed at national and international events and a lighter, more colourful and sensuous architecture, although this was not perhaps the most important feature. The particular achievement has been the highly successful integration of these projects into the existing city centre fabric. Imaginative urban design was based around two principles. The first concerned the use of industrial heritage in the form of the eighteenth-century canal network as both a focus for development and a pedestrian circulation system. The canal was Birmingham's equivalent of a waterfront and has been exploited to high effect. The second element was the improvement and articulation of urban spaces not just within the ICC complex itself but in its connection to the rest of the city centre. A carefully designed sequence of urban spaces was built (linking Centenary Square, the atrium of the brutalist Public Library, Victoria Square and pedestrianized New Street) involving the sinking of the inner ring road and first phase of the breaking of the 'concrete collar' which was strangling geographical expansion and spatial improvement of the city centre. European Structural Funds were again critical in financing this urban design breakthrough. Private sector investment followed in Brindley Place on a 7.5 ha site directly across the canal from the ICC, originally conceived as an American-style festival marketplace along the Baltimore model but subsequently developed as a cold, controlled office complex with some entertainment facilities and gated housing along the canal-side.

5.2 One side of the International Convention Centre connects to Birmingham's 'waterfront' – the canal system – combining natural elements and built heritage in the reconstruction of the city's image.

The ICC complex was not without controversy. Development between 1986 and 1992 involved some £380 million of public spending, including external and especially European Structural Fund sources. This represented 30 per cent of the city's entire capital programme, whilst £123 million less was being spent on housing than the average local authority nationally, capital resources devoted to education were falling by 60 per cent and revenue expenditures were also being diverted away from mainstream social programmes to cover interest charges, for example £51 million from education revenue support between 1991–92 (Griffiths, 1998, based on Loftman and Nevin, 1992; 1994; see also Newman and Thornley, 1996; Barber, 2001). Birmingham invested more heavily in city centre redevelopment than similar UK cities and, allegedly, at considerable cost to the city's social programmes and deprived population. This was sufficient to provoke a change in council leadership in 1994, with Theresa Stewart coming from outside the close-knit Labour group driving the ICC initiative to lead public expenditure back to 'front line services' of housing, education and health.

Nevertheless, the recycling of the city centre continued, though now more reliant on private sector and external funding. The Highbury Initiative strategy of redevelopment based on specialized quarters was pursued steadily and self-consciously, on occasion based on local tradition as in the case of the Jewellery Quarter, at other times based on inventions such as the Chinese Quarter. Office development took place on a modest scale in the existing sector, transport infra-structure was improved and street design and pedestrianization was extended. More recent developments in the new millennium have seen a revival of large-scale projects such as the further destruction of the 'concrete collar' inner ring road at Masshouse, the redevelopment of the Bull Ring shopping centre and a major regen-eration programme for the Eastside district to the west of the city centre. The claim of city centre recycling is not therefore an exaggeration but its sense of sustainability is less clear. Some of the social implications insinuated in relation to the ICC project demand further examination.

Environmental consumption and gentrification

As with other cities which have embarked on city centre regeneration, the Birmingham experience has been open to criticism for its elitist character, aimed principally at the affluent middle class and high-earning young. The very nature of its rebuilt economy gave priority to professionals in financial and business services, and small enterprises in the skilled craft and creative sector industries which demand considerable cultural capital. New civic facilities are mainly the 'high arts' of theatre, classical music, museums and exhibitions. The less traditional side of contemporary cultural expression and the entertainment industry, such as 24-hour night-life, alter-native living, gay quarters, ethnic celebrations and so forth have not taken off as in comparable UK cities such as Manchester and Glasgow. Similarly, leading new shop-ping facilities in the city centre are decidedly mid- to upper-range of the market, as is new housing provision. New housing in the city centre was slow to take off, with the first significant development occurring in Symphony Court, the residential component of Brindley Place. Its canal-side, gated, high-security environment set the tone for subsequent developments concentrating along the Broad Street corridor and in the Jewellery Quarter. In 2001 it was estimated that some 1600 city centre new homes had been built with a further 1800 in the planning pipeline. They are all in the upper end of the market, and include corporate homes, second homes for affluent buyers and investment opportunities for the rental market (Barber, 2001).

In short, city centre environmentalism contains a set of economic and symbolic filters which effectively exclude the poor, the old and some ethnic groups. When these groups do participate, as low-paid workers or occasional visitors, the surveillance is heavy; Birmingham certainly *feels* like one of the most CCTV-intensive cities in the country. The exclusionary nature of city centre development was especially sensitive in Birmingham, given that public investment there had been higher than in other cities and the social costs more politically contested. The gentrification of the city centre is well-portrayed in the following description of a scene from the office/restaurant edge of the ICC convention quarter:

> Brindley Place, Birmingham, 2001. Canalside café terraces bustle with conversation and laughter in the fading hours of a bright late spring day. Medical consultants from across Europe discuss the day's proceedings at their biannual conference, property professionals unwind after a hectic week of meetings and deals, while in the adjacent public square a group of models and PR executives wrap up a photo shoot for a local lifestyle magazine. Middle-aged classical music buffs trickle into restaurants for some sustenance before heading on to the evening concert.
>
> (Barber, 2001: 6)

Cameos of this sort can now be found across the city centre. The new Bull Ring shopping centre, opened in 2003, provides a retailing equivalent of Brindley Place at the other side of the centre, as well as the opportunity for Birmingham to make a competitive mark in its otherwise largely unspectacular architectural achievement. Architecture critic Deyan Sudjic (2003) waxes lyrical: 'Birmingham's Bull Ring shopping centre is architecture's version of the demilitarised zone that divides North and South Korea, the site of a non-meeting of minds on a truly epic scale, where two different views of what the world is about come within inches, but fail to acknowledge each other's existence'. Just when you think he must be referring to the extravagant building's relation to the run-down Digbeth area next door, it becomes clear he's simply contrasting two parts of the new shopping centre itself, as he continues: 'The boxes stop, and equally abruptly Selfridges, with its dizzying skin of 15,000 polished aluminium discs, comes billowing out like a giant blue and silver cloud'. Nevertheless, the two-worlds analogy and architectural metaphor can usefully be extended to the city as a whole. The exclusive and expensive Selfridge's store, a highly-publicized coup for Birmingham's retail and marketing strategy, does indeed seem to symbolize the chasms in the city's social life and expectations, whilst its curvaceous form, inspired by a Paco Rabanne evening dress, attempts the definitive statement that dour old Birmingham is dead. Flamboyant consumption is its message, just as flamboyant culture is the message of Bilbao's Guggenheim Museum or flamboyant government that of Edinburgh's new parliament building.

However, the new Bull Ring is also another clever piece of urban design, with a small artificial rise in the street level entrance producing some notable townscape effects, followed by a descent through formal public spaces into the revitalized market area. Modest pensioners, white and black, do sit out on the edges on a sunny afternoon, as do poor kids skateboard in Victoria (but not Centenary) Square over towards the ICC. But they seem more transient and obtrusive, less part of the new city centre environment. In the mid-1990s the following comment was probably typical of the early perceptions of the ICC of many social groups:

> I haven't been to the city centre for years, but I went to see this new stuff going up around Broad Street. It's definitely something you can point to, show friends and family, say that it makes you proud to be a Brummie. Having said that, I'll probably never visit again – I don't really know who will, it's not really aimed at locals is it?
>
> (cited in Griffiths, 1998: 54)

Perceptions of this sort were behind Birmingham's turn to 'internal marketing' to try to make ordinary people feel involved and part of urban restructuring. The visit by USA President Bill Clinton was one of the first major 'hearts and minds' campaigns, with its well-known 'Hold your Head Up High' campaign involving posters with local residents and phrases such as 'Get your hair done, Marge. Bill Clinton's coming to town'. Later, the same Birmingham Marketing Partnership ran the 'Be a visitor in your own city' campaign for residents hosting friends and family (Barber, 2001) and, for the European Cultural Capital competition run in 2003, the more quirkily ontological 'Be in Birmingham'. Other attempts to involve the ordinary citizen were made at the Highbury 3 conference in 2001, where it was discussed how to connect 'communities' to Birmingham's new spatial economy developing around the city centre and the subsequent creation by the regional Advantage West Midlands development agency of two regeneration zones on major axes connecting to the international airport and to south-west of the conurbation.

Economic restructuring and the greening of poverty

The preceding description of the socio-spatial effects of the privatization of the city centre environment is not dissimilar to the widely analysed dynamics of most large West European cities. In its own terms it has been relatively successful as an economic strategy, contains some notable urban design and is generally agreed to have marked a vital change in the city's self-image and national and international status. Given the paucity of natural features such as water or striking landforms and a generally low-grade physical environment, this achievement is even more remarkable.

However, it is also widely acknowledged that not only is the city centre environment (for that is how it is described) elitist and exclusionary in itself, but has also been reconstructed at the cost of the rest of the city, especially the poorer sectors.

During the early stages of economic restructuring, the ICC complex was criticized not only in terms of the scale and social implications of the city's financial commitment, but also for its limited impact on direct and indirect employment generation, inward investment attraction and job benefits for the city's deprived populations (Loftman and Nevin, 1994, 1996). Concerns over long-term employment prospects were common in large cities developing a service economy where jobs are typically low-paid, unstable and with poor career prospects. In Birmingham this was particularly problematic in view of the limited 'cultural capital' and the ethnic diversity of the city's deprived young population and the traditional manufacturing workers' lack of 'interpersonal skills' required by the leisure industry. Conditions were made especially precarious in the context of the UK's notoriously deregulated labour market. Those early fears surrounding the ICC complex seem to be borne out by later employment trends in the convention quarter. The Brindley Place office development was providing over 3700 jobs in the telecommunications, financial, legal and professional service sectors by 2001, with only some 1360 new jobs in the leisure industry in the surrounding Broad Street corridor (Barber, 2001). Total employment in Birmingham's financial, professional and business services sector grew from approximately 79,000 to 102,000 between 1991–2001, projected to rise to 122,000 by 2010.

Such employment trends in economic restructuring have a clear spatial dimension for the city as a whole. Whilst the long-term decline in the city's total employment was reversed in the mid-1990s, Birmingham's unemployment still stood at 7.7 per cent in early 2004, compared to a national rate of 2.5 per cent. The traditional manufacturing sector is expected to continue to decline, with its contribution to total employment falling to under 20 per cent, whilst overall investment in the urban economy has been modest (Barber, 2001). For its part, employment provided by the new service economy is concentrated in high-wage professional services and low-wage unskilled, unstable and part-time jobs. This leads inevitably to greater socio-spatial segregation, and areas with ageing populations, low educational levels and/or high black and ethnic minority populations became concentrations of multiple deprivation. For example, in early 2004, unemployment in six of the city's 39 wards exceeded 15 per cent, more than twice the city's average (and six times the national rate), with a high incidence of low-paid and part-time work (based on Birmingham City Council, 2003).

These and similarly disadvantaged sectors of the city were precisely the areas targeted by the city's regeneration and New Deal for the Community programmes that the new economy was failing to reach. As we saw earlier, the main thrust of regeneration was to get people into the labour market through skills training involving multi-agency support and community commitment. However, the environment was also a key aspect of regeneration in these areas. Regeneration involved not only housing improvement and campaigns against crime and anti-social behaviour but also attempts to improve the physical environment through reclaiming derelict land and buildings, tree planting, park and open space maintenance, traffic calming, and so on, as well as a gigantic landscaping spree along the major roads.

5.3 All major roads in Birmingham have been heavily landscaped in an attempt to 'green over' the city's modernist excesses of the recent past.

5.4 Extending uban envronmentalism to the poorer neighbourhoods, with some modest traffic calming and landscaping.

Without suggesting that this was the conscious intention, the overall result has been the draping of urban poverty with swathes of greenery, with every available inch of public open space employed to conceal and 'naturalize' social inequality. The logics behind the greening of poverty lie much deeper. They are to be found in the discursive construction of the environment as the definitive area of urban quality and key to economic recovery, as established in city centre redevelopment, and the simulation of environmental interventions in the poor sectors. The environment became a form of social integration in the rebuilding of the city, whereby those sectors excluded from the material benefits of the new city centre economy could participate, from the margins, in its symbolic spatial manifestations. With the more recent sustainability policy, the environment has also been established as an area of civic duty, community solidarity, personal responsibility and for the regulation of citizen behaviour.

Spatial organization and the ecological environment

In recent years spatial planning at the regional level has gained importance in the UK. Regional Planning Guidance for the West Midlands region is the obligatory framework for the region's economic, transport and sustainability strategies, major land use decisions and local authority planning. One of its explicit aims is to promote sustainable development, and in this direction seeks to concentrate development in the major urban areas, promote urban renaissance and modernize transport infrastructure. This spatial strategy follows the idea of the compact city-region with its aims to improve public transport and reduce the need to travel by making existing settlements more equitable, diverse and attractive. However, as Murie *et al.* (2003) have observed with regard to the West Midlands region, whilst individual local authorities may actively support regional spatial planning and its sustainability ideals, in reality this is severely hampered by central government's control over many regional planning agencies, major land use decisions and the funding of transport and infrastructural projects. Furthermore, the priority given to market solutions and business interests creates a serious obstacle to equity issues at the spatial level. The underfunding of major infrastructure has long been a chronic problem in the UK.

Evaluation of progress towards urban sustainability made in the Birmingham Local Action 21: Achievements Local Agenda 21 1992–2002 and the Sustainability Strategy and Action Plan 2000–2005, highlight achievements in terms of city centre redevelopment (pedestrianization, recuperation of public urban space, introduction of housing, creation of socially sustainable quarters, recycling of old office blocks) and advances in public transport (Metro Line One Tram, better bus services with low emission engines, low-level boarding, bus lanes and high-quality bus stops). Negative counter-trends concern the inability to intervene in the housing market and reverse socio-spatial polarization, growing car ownership and use, delays in the extension of the Metro light railway system and the opening of a privately financed motorway toll road, and city centre redevelopment based on an economy of unlimited and sumptuous consumption. These are, of course, the spatial manifestations of the structural contradictions of advanced capitalism, neoliberal government and an individualized consumer society. The pursuit of sustainability within these conditions is largely limited to good intentions and the monitoring of self-set goals through performance indicators.

The City Council acknowledges the shortcomings of certain measuring and monitoring procedures but gives the overall impression that the city's natural resource systems are in an acceptable condition and that pollution is no longer the major issue, although increasing car use is a concern. Even with regard to air quality, pollution levels are shown to fall within objectives for all major pollutants, except for nitrogen dioxide concentrations along the high-volume traffic corridors (West Midlands Chief Officers Joint Working Group, 2000) and pollution is predicted to remain within objectives in the short term. Noise is declared not to be a major problem and there are no registered sites of severe land contamination. As for biodiversity, the nature conservation strategy has been operating since the early 1990s, and the city might reasonably claim acceptable levels of natural open space/wildlife habitats within a once highly industrialized area.

The Environment Agency Midlands Region (2004) identifies the main regional environmental problems in terms of pollution as more cars take to the road, flooding due to global warming, growing demands on water, land contamination and waste management. The Agency makes a general call for greater efficiency in resource use and the control of demand, and Birmingham City Council has assumed the role of example-setting through such things as internal environmental auditing, encouraging employees to use public transport and green purchasing and fair trade policies. However, as with most big cities in the UK, demand-side policies have meet with limited success. Car journeys, the biggest source of pollution, continue to increase and the city aims to halve projected levels but still only maintain traffic growth *increases* within 5–8 per cent over the period 1996–2011 (West Midlands Joint Committee, 2003) Only 6.5 per cent of domestic waste was recycled in Birmingham in 1997–98, while disposal is mainly through incineration, with a programme of heat generation under consideration. The reuse of bottom ash from the waste plant, a garden waste composting scheme and the introduction of 'wheelie bins' should improve the situation, although it is doubtful that the city will reach the central government recycling target of 25 per cent for 2005 (Birmingham City Council, 2000b).

The city presents a picture of improvement in the state of natural resource systems. Birmingham's sustainability monitoring, introduced in 1997 through the design of 22 indicators, showed short-term improvements in air quality, the number of businesses working towards recognized environmental standards, the percentage of open water of good quality, the percentage of domestic waste recycled, more local exchange trading schemes (along with more under-5s in nursery schools, and falling unemployment and suicide rates). On the other hand, there were negative trends in passenger miles travelled, a lower percentage of the population living within half a mile of local amenities, greater fear of crime, lower participation in local elections, a higher incidence of heart disease fatalities, and more miles travelled by selected food items consumed in Birmingham.

Both global ecological and local social issues focus on citizens and sustainability policy urges people to 'do their bit' in their homes (energy conservation and waste separation), how they travel (more walking and use of public transport) and what they consume (dietary and tobacco control). They implore healthy lifestyles for a healthy planet, whilst systematically avoiding the material conditions within which people's lives are played out and the unhealthiness of the demanding routines of work, the effects of poverty and local injustices of social inequality. The notion of environment subsumes these diverse phenomena into a single whole, at the same time both altruistic and life threatening. The local Friends of the Earth branch (2003) seems to go along with this overall agenda, aiming criticism primarily at major transport and land use issues, whilst technological solutions are supported by the business community. However, citizen resistance to sustainability programmes is widespread, whether in the form of cynicism, egoism or disinterest, a resistance the city authorities try to erode through the cultivation of fear, especially of crime but also of floods, ill-health and sanctions of various kinds.

Conclusions

The central theme emerging from the analysis of environmentalism in Birmingham is its inextricable relationship with economic restructuring and property-led urban development. The notion of environment was discursively constructed around city re-imaging through architecture in the mid-1980s, and this is still the case 20 years later. Architecture and urban space continue to dominate in this aestheticization of the environment and the private sector maintains its strategic control over urban environmental discourse. However, the scope of environmentalism has widened significantly in recent years, as area-based regeneration programmes have extended environmental discourse to all parts of the city and the 'community'. In the process, the environment has been employed not just to stimulate investment and the new service sector economy but also to encourage citizens to adapt to the work requirements and civic responsibilities of the post-industrial, entrepreneurial city. In short, the environment has been constructed as an integral part of the restructuring of capital and the re-regulation of labour in Birmingham's transforming economy.

The key concept has been the notion of quality. Environmental quality was consistently argued as the prime requirement of competitiveness and the ultimate measure of the everyday urban experience of citizens. This was no casual association of ideas. A complex coalition of institutions ensured that environmental quality became the definitive feature of urban development aspirations, urban management goals and urban evaluation criteria. What at first sight appears to be a remarkable return to environmental determinism, given legitimacy first through a discourse of survival and later through the social content of sustainability, was, of course, structurally integrated into urban development dynamics. The notion of environmental quality in Birmingham was constructed in the very specific and demanding circumstances of long-term industrial decline and the obligation to restructure the city's economy in the context of new globalizing dynamics and urban competition. The initial phase of environmental discourse, argumentatively directed by and in the interest of capital, was undertaken in the period of radical neoliberalism of the 1980s whilst the shift to community, the labour market and citizen subjectivity unfolded in the re-regulatory phase of the mid-1990s onwards. Several important aspects of this process can be drawn out.

First, the environment was not simply a glossy façade to post-industrialism but an integral part of the spatial restructuring of the city's economy. The crisis of Birmingham's old Fordist-style engineering and automobile industry resulted in the city's post-war modernist spatial organization and urban aesthetic becoming as redundant as large parts of the traditional economic base. The environment became both the opportunity for new capital circulation in the property market and the spatial requirement for the development of the service sector and the cultural economy. New forms of governance, partnership with the private sector and institutional innovation, complexity and obfuscation were vital in delivering both environmental meanings and physical realities.

Second, the construction of the environment and the emergence of sustainability policy was not a rational progression towards ecological enlightenment but part of the strategic exercise of political and economic power within the city. Environmental meanings were constructed in a period of radical social change and conflict, where the privileging of economic interests was obscured by the symbolic function of the environment in creating a sense of city pride and inclusion, reshaping social aspirations and citizen subjectivities, regulating behaviour and establishing new moral obligations in a privatized social landscape. The highly political nature of the environment-as-development became neutralized as a common-sense requirement (of urban quality) and collective global responsibility; in this sense the environment fulfilled an important ideological function.

Third, the reality of environmentalism in Birmingham was far removed from the holistic, integrative quality of the environment as an abstract concept. Urban development discourse assured that the environment was disassembled and reconstructed in order to facilitate and legitimate differential spatial appropriation, in a subtle, discursive distribution of rights of access to, control over and responsibility for different parts of the city. The integrative possibilities of the environment were limited to the symbolic realm, the reach of which has been limited by the manufacturing tradition and ethnic diversity of the city's population.

Finally, the rationale of ecology has played an insignificant role in Birmingham, even when introduced at a late stage within the framework of sustainable development. The inclusion of references to global ecological problems was simultaneously translated into indicators of social behaviour, calibrated according to the city's political necessities and economic development possibilities. In this way, the environment was employed to both create and conceal the logical and social contradictions of its sustainable development policy – for example, appeals for the prudent use of natural resources within an urban policy based on intensified consumption and social inclusion amidst growing inequality and polarization.

Chapter 6
Urban environmentalism in action 2:
democracy and participation in Lodz, Poland

Introduction

Lodz is a city of contrasts and enormous recent changes which challenge a conventional linear reading of space, place and the environment. It is simultaneously provincial and outward-looking, solidly socialist and loudly entrepreneurial, sedate and brashly postmodern. The two-hour journey by road from Warsaw begins on a fast dual-carriageway but somewhere in the middle slows down to a gentle ride over undulating countryside and through small quiet towns. The central bus and rail station is almost quaintly dilapidated. The surrounding old town is four-storey, grey and spacious, suggesting a disciplined opulence now disappeared, a cityscape occupied by what seem to be mostly solitary individuals with dogs, made small yet dignified by an urban scale of an unhurried, half-empty, quiet magnificence not quite realized. This sensation is soon shattered. The single main street is a brightly-lit buzz of youthful commerce which terminates at one end in an American-style shopping strip. The middle ring-road is a congested circle of hypermarkets and new industrial premises surrounded by the mass-housing estates and bucolic allotments of the socialist period. Further out still are new residential districts of the *nouveau riche*. Where not long ago the second language was Russian, many people on the street now have a smattering of English.

The city is situated in the centre of Poland some 140 kilometres to the west of Warsaw. It is the smallest of the three cities covered in this book, with a population of just under 800,000 and centre of the Lodz province or 'voivodeship' consisting mainly of smaller industrial towns. Although its early origins date from the late mediaeval period, receiving city status in 1423, it is fundamentally a product of nineteenth-century industrialization. Situated on an upland ridge between the country's two major rivers – Vistula and Odra – the plentiful streams provided abundant clean water and power for the flourishing woollen, linen and cotton mills that sprung up from the beginning of that period.

Equally important is the geopolitical history of Lodz. Throughout the nineteenth century, indeed right up to the present day, Lodz's fortunes have wavered in the wars and alliances of empires and régimes of East and West (Liszewski, 1997). After the second partition of Poland in 1793, Lodz came under Prussian control, then Napoleonic influence saw the creation of the Duchy of Warsaw and later the Congress Kingdom of Poland (following the Congress of Vienna alliance with Russia), a period of relative independence which saw the beginnings of the textile industry. Between 1820 and 1830 a completely new city emerged, when the current main shopping street was laid out in its original form as a linear arrangement of bourgeois mansions and institutions backed by mills. Russian reprisals following the 1830 Uprising resulted in import duties on manufactured goods, but technological advance continued and when restrictions were lifted in 1851, the textile industry took off. Lodz grew from a small town of 15,000 inhabitants to a large city of nearly half a million by 1914. Trade with Russia intensified and in the second half of the nineteenth century 70 per cent of textile production was sold to Russia and Lodz concentrated some 30 per cent of total industry in the Kingdom of Poland and 50 per cent of textile production.

Events of World War I had disastrous consequences for the city. Germany confiscated industrial machinery and finance capital, and securities in Russia were lost to the Soviet Revolution of 1917. Industry and the local economy slumped, barely compensated by the location of administrative functions under the independent Polish state. Then World War II saw German occupation, a war-oriented economy, severe damage to industrial infrastructure, the extermination of the Jewish community and German withdrawal, though not extensive physical damage to the city itself. The post-war settlement saw Poland assigned to the Soviet bloc, socialism and a centrally-planned economy. Some industrial diversification occurred and a new spatial structure arose around the old nineteenth-century city. The implosion of the Soviet Union was to witness a return to liberal democracy after 1989, the beginning of a period of system change in the transition back to capitalism and the eagerly embraced challenges of a market economy, Europe and globalization.

It is in the period immediately before and during transition that we will examine urban environmentalism. Not only is Lodz a fascinating case of fluctuating urban dynamics brought about by geopolitics, it also provides the opportunity to explore the ways in which radical changes in the conception of the environment, political practices, development processes and management forms have constructed environmental meanings and urban spatial realities, all modulated by more permanent cultural beliefs and traditions. With recent Polish admission to the European Union, in what way might Lodz be considered a 'sustainable' city?

Urban discourse: the environment in transition

Tracing the role of the environment in the urban development of Lodz is complex and cannot be captured fully in the policy field alone. Whilst due attention needs to be given to policy discourse, it is also apparent that in a transition country such as Poland changes have gone far beyond the scope of policy and even politics in a conventional understanding of the terms. General process models of politics/policy, whereby policy follows from the politicization of issues into affairs which demand public intervention, are clearly inadequate. Evidently, system change has been so momentous that the incremental assumptions behind such an approach fail to capture the scale and scope of transition from state socialism to globalized capitalism, and the particular challenges and dilemmas this implies for urban development.

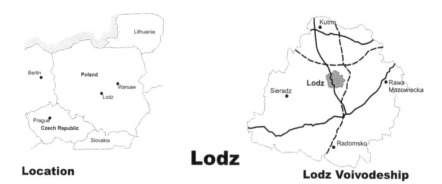

Location

Lodz

Lodz Voivodeship

■ Historic centre
▨ Mixed use/style housing
▨ Large-scale socialist housing
▢ Recent suburban growth
■ Major parks and woodlands

6.1 Map of Lodz.

Transition or the 'velvet revolution' in Central and Eastern Europe was no less profound for the non-violent nature of its unfolding. Not only did political and economic régimes change, from a centrally-planned economy overseen by the Communist Party to a market-oriented liberal democracy. With it came a change in ideological understandings, the institutional organization of society and the shape of everyday experience. A new set of assumptions concerning development imposed itself over a declining socialism, fresh social dynamics were set in motion, individual lives recast. All of these had profound implications for ways of understanding and interacting with nature and the construction of meanings of the environment. However, as the term implies, transition involved a gradual process laden with past ideas, values and practices. In Poland in particular there was no clean break with socialism, no great symbolic moment of change or monumental occasion to celebrate a sense of revolutionary transformation. For a long time most people's everyday lives continued much as before, but more difficult and uncertain in outlook. Old political and social orders resisted and adapted to new capitalist realities. Traditions from the past re-emerged alongside the appearance of new horizons for the future.

Nevertheless, fundamental changes did occur, with regard to the environment as in all spheres of life. In particular, transition towards a capitalist economy involved a different way of understanding nature and the environment, and their significance in terms of the urban development process. Underlying these changes was a new epistemology or way of understanding the natural world. At an official level, Marxist–Leninist political philosophy and scientific socialism gave way to the assumptions of liberal democracy and individual freedoms. The narrative of nature and the environment was changed, along with the theoretical explanation of the environmental problem. New concepts and story-lines of development were introduced, which in turn shaped and legitimated policies and development processes. The rationale of environmental administration, the institutional circulation of information and voices of authority became resituated. However, it was a complex process with all the inconsistencies of any discursive formation made doubly evident in the paradigm change of transition.

The understanding of urban environmentalism in Lodz therefore requires a wide analytic perspective. It is not just that discourse subtly moulded meanings to specific political ends; the political system itself changed. In terms of discourse analysis, a new discursive formation arose to supplant the meta-narrative of socialism. The character and urban implications of such change are explored in the following section.

The concept of nature and the environment: the communist version

The communist period inherited a polluted Lodz. It was an industrial city suffering from heavy air pollution arising mainly from the use of low-grade coal for energy production and heating. Much of its blackened physical environment was run-down, unhealthy and obsolete, with inadequate sewage treatment producing heavily contaminated rivers. It was, in short, typical of most industrial cities in both Eastern and

Western Europe in the immediate post-war period. Some 40 years later, one of the features of Western triumphalism after the fall of the Soviet bloc was the exposure of the scale and extent of continued environmental degradation under communism. By the end of the 1980s, cities in the West had cleaned up considerably and the new green urban agenda was taking shape. The capitalist world was setting the conceptual ground for sustainable development (ecological modernization) and beginning the wholesale export of its dirtier industries to the South, whilst almost revelling in the environmental mess uncovered in the Soviet bloc countries. The picture was far from uniform between countries, but the overall situation was indeed disturbing, with widespread industrial pollution, hazardous wastes and under-maintained plant, with the Chernobyl nuclear disaster at the forefront of the public imagination.

How had this situation come about? It was certainly contrary to Marx's theory on the relation with and control of nature. Given the centrality of Marx to the epistemology of the environment, certain aspects may be briefly recapitulated. As we saw in Chapter 4, the early philosophical works of the young Marx insisted on humankind's integration with nature – our 'species being'– describing nature in terms of humankind's 'inorganic body', the source and condition of human existence. His idea of the conscious action of humankind on nature resulting in the 'humanization of nature and the naturalization of man' is as good as any definition of the production of the environment. At the same time, Marx's dialectics of nature emphasized the conscious, practical character of this interaction, in which humankind distances itself from nature without ever being able to extricate itself fully from a dependence on nature and the biological fact of human existence. In his later works, Marx explored the historical nature of humankind's relation with nature, particularly under capitalism. He coined the term of an 'organic exchange' with nature to accommodate the idea of a mediation of this relation through the social relations of production. Under capitalism, argued Marx, these relations become sufficiently distorted to give the impression of an independence from nature except in a general, abstract sense. Through commodity production, nature ceases to be recognized as a force in itself and becomes a mere object of exploitation. From a Marxist viewpoint, only the rational control of the interaction with nature through scientific understanding and production aimed at the satisfaction of human needs could avert an environmental crisis.

Thus it was not so much that, under the labour theory of value, nature had no value in itself, rather that this value was lost from sight under commodity production or the production of exchange value. Nature, insisted Marx, was the source of wealth, along with the labour invested in its transformation. The secret lay in the rational, unreified relation with nature and its forces. Inspired by scientific discoveries of the nineteenth century concerning the fluidity and movement of nature, the power of technology and the potential of humankind to rationally exploit the dormant features of nature for its own improvement, Marx implied that the major obstacle was capitalist obfuscation and alienation.

Much academic debate has since taken place concerning the Marxist basis for understanding the environmental problematic. Critics were quick to denounce

Marx's conception of nature as suffering the same limitations as Western capitalist thought and the deficiencies of the Enlightenment project: objectivation, narrow scientific understanding, the pretence of dominance, the fallacy of progress, the rational exploitation of an infinite resource to be incessantly exploited through the application of technology. Others argued that Marx failed to appreciate the limits of nature and therefore foresee the global environmental crisis or the 'second contradiction' of capitalism.

Soviet scientists were happy to accept Western Marxists' opinion of the incompatibility between capitalism and ecology, but refuted that this was an unforeseen problem. For them, the environmental problem arose from the character of that relation under capitalism. Commodity production and the profit principle led inexorably to the exploitation of both workers and nature, spiralling out of control in the expansive dynamics of capitalism. The answer lay in replacing the destructive chaos of capitalism by the scientifically-based and rationally-planned exploitation of nature. Reason was to triumph over competition and the profit motive as the ordering principle of society and its relations with nature. Soviet scientific orthodoxy held the view that:

> The theoretical analysis of the ecological problem from the perspective of historical materialism leads to the inevitable conclusion that elimination of ecological danger is only possible through radical social transformation of the capitalist world, the suppression of private property and class antagonism. Only in this way can positive technological methods be applied, in a universal way, to totally overcome the antagonistic ecological contradictions which are inherent in the scientific-technical revolution in the context of the capitalist system.
>
> (Frolov, 1981: 20, our translation)

Neither were they afraid, in contrast to Western environmentalists, to insist unhesitatingly on the notion of progress:

> The dialectic approach implies renouncing the metaphysical separation and contraposition [under capitalism] not only of society and nature, but also of the mode of production, 'technics' and technology from the socio-economic forms in which they currently exist. For this reason, the dialectic approach rejects contemporary answers to ecological questions which are limited to the weighing up and choice of technological means ... It is also becoming increasingly understood that it is impossible to maintain the 'natural equilibrium' of biospherical processes through conservation. The constant transformation of nature and optimization of the biosphere through continued scientific and technological progress is precisely that which will permit the achievement of harmony between man and nature.
>
> (Frolov, 1981: 15, our translation)

It may be assumed, then, that the rational control of nature based on scientific understanding and the application of technology, exercised through the centralized planning of a socialist economy, were the major official themes of environmental planning in post-war Poland. However, by the end of the socialist period Lodz, like many industrial cities of the Soviet bloc, was not only some distance from achieving harmony with nature, but apparently had achieved less than most cities in the capitalist West. Or had it? Certainly the urban environment was generally more run-down and natural resource systems more polluted. Nonetheless, comparisons with the West were made on the basis of the assumptions and aims of an environmental agenda which privileged the protection of the environment conceived as an external physical entity, the relation with which was increasingly determined as an area of consumption; in other words, precisely in those terms criticized, at a conceptual level, by Marxist theorists and socialist ideology. If Western cities had become noticeably cleaner, greener and less contaminated, they certainly were not in greater harmony with nature in the Marxist sense. The mediation of the relation with nature had intensified under the extension of the commodity form, to such an extent that any real sense of original contact with nature had virtually disappeared in the West. Nature had been reconstructed and inserted into the city as a social product – waterfront developments, urban villages and nature parks, organic farm produce, and so forth – whilst the global reach of urban consumption was seriously damaging the environment at a planetary level.

On the other hand, the objective conditions of the physical environment were undoubtedly deficient in a city such as Lodz. 'Really existing socialism', just as 'really existing capitalism', was never quite up to what it claimed to be. The purity of theoretical positions was unable to impose itself completely on the complexities of reality. In the case of the socialist bloc countries, the Stalinist drive for industrialization, the overriding aim of creating a working class proletariat and then the Cold War became political priorities which created an implacable distance between scientific socialist environmentalism and communist development strategies. After World War II, Soviet influence carried these epistemological inconsistencies and political conflicts into Poland. Far from being abstract considerations, they helped shape Polish cities in a decisive way.

Urbanization and spatial development in the socialist period

The urban implications of the strains between scientific environmentalism and really existing socialism are well illustrated in the case of the development of Lodz in the communist period. The city itself was not heavily damaged during World War II, although the social consequences were enormous, with the Jewish community annihilated and the German elite withdrawing overnight. A now ethnically homogenous Lodz came under the influence of the Soviet régime and its economic policy, with important consequences for urban development. Lodz, already a city with a long industrial past, was systematically rendered a workers' city and subject to policies which progressively emphasized its proletarian character.

There are several general characteristics of post-war urbanization which should be mentioned. First, urban development was led by the national policy of socialist industrialization (Wodz and Wodz, 1998). The creation of the Polish People's Republic in 1949 implied its acceptance of an economic development policy based on heavy industry and the expansion of power and energy sources. Industrial location was aimed at spreading industry as evenly as possible, usually around the larger cities or via the creation of new towns. In the case of Lodz, this meant a consolidation of the traditional textile base and some widening to include engineering and chemicals. Industry-led urbanization meant that in Poland as a whole there was massive shift of population from rural to urban areas. Between 1950 and 1990, the number of cities of over 200,000 inhabitants increased from 7 to 20, and their share of the total population rose from 25 to 50 per cent. Lodz itself grew relatively slowly. For some years after the war it was prohibited to move to the city in order to hold back the intense migration pressures from other war-damaged regions, and smaller settlements in the region absorbed much of the population growth. The destruction of Warsaw led to Lodz for a while acting as the informal capital, a situation that stimulated the institutional development of the city (Liszewski, 1997).

Second, industrialization meant the mass construction of housing to accommodate the new urban proletariat. Industrial building systems developed in the Soviet Union were imported into Poland and used to meet the huge housing demand. During the early years these systems were used on a relatively small-scale basis, and it was only in the 1960s that massive projects were developed in Lodz. Projects in this period were not only large scale, but were also built to lower space and construction standards. Third, the predominance of industry in the urbanization process meant that the city in general was conceived as a production unit which should operate with functional efficiency. This led to a general imbalance between

6.2 Housing from the socialist period with combined road and tram systems. This is the largest in Lodz; most housing projects were smaller in scale and often arcadian.

industrial investment and the production of consumer goods and investment in technical and social infrastructure of cities (Szymanska and Matczak, 2002). Fourth, socialist urbanization policy brought with it the subordination of local development to national policy, and of local government to the party apparatus. Spatial development plans tended to be mere reflections of national economic strategy, over which the city's representatives had little or no influence. The role of city authorities tended to be limited to the execution of national policy as determined at a local level by the party machinery and production managers. This in turn did little to stimulate the development of an effective civil society.

What resulted in spatial terms was a generalized picture of urban and architectural modernism. However, the reality in Lodz was not as monotonous and uniform as might be imagined. As Kaczmarek (1997) has pointed out, Lodz is in effect a mosaic of different architectural styles and spatial configurations. The nineteenth-century town remained largely intact, much of the early state housing was small scale, old public buildings remained and new ones were built, presocialist villas survived, industry was redirected towards zoned areas, and so forth, and all of this was interspersed with ample open space, parks and allotments. This spatial diversity was matched in social terms. The new urban proletariat was largely of peasant origin, keeping alive traditions of religion and the family and, more importantly in Lodz, neither private enterprise nor civil society completely disappeared in the process (Szul and Mync, 1997). By the end of the 1980s Lodz, though a heavily industrial city, was far from the dull production unit of socialist development policy. In fact, neither was it that efficient in production terms, as investment had been lacking not only in social infrastructure but also in industrial technology.

In environmental terms, the industrial impulse of socialist development solved some problems and created others. Despite the shortcomings of industrial systems, housing conditions improved enormously for the mass of the population. Modern accommodation with basic services and central heating meant a vast improvement in comparison to the nineteenth-century stock of tenements and rural dwellings, and as social units they still work relatively well. The demise of such developments in Western cities, where similar estates became 'sinks' for the 'socially inadequate', was due not so much to inherent architectural shortcomings as the social changes brought about by the structure and dynamics of post-Fordist development. Architectural modernism and its aims became incompatible with flexible production and societal individualization, in which property ownership became a cultural imperative and a key aspect of capital accumulation.

The socialist city also had plentiful open space and extensive and environmentally clean public transport. However, air and water pollution remained serious problems. Investment in public health infrastructure was not a priority, industry lacked the investment potential and incentives to develop cleaner technology, and the dependence on high-sulphur coal continued. Smoke-chimneys remained a powerful symbol of work and progress for the majority of the urban proletariat whose living conditions improved considerably in Lodz. There was little public concern for environmental issues until well into the 1970s, a fact aided by the suppression of environmental information for political convenience. The epistemological basis of environmental thought and policy were impeccable, but not the reality of urban management. This is explained in part by the fact that socialist urbanization was an essentially political project, a strategic element in the creation of an industrial proletariat and a workers' state, especially evident in the case of more bourgeois cities such as Krakow where industrial expansion was used as a means of undermining the influence of independent intellectuals. Ironically, scientists would eventually play an important role in developing environmental awareness and oppositional politics, and Krakow itself became the birthplace of the influential Polish Ecological Club. However, before moving on to environmental politics in the transition period, we shall take a closer look at environmental policy in the socialist period.

Environmental policy under socialism

As indicated earlier, the assumption of a *tabula rasa* existing at the beginning of transition would be untrue. Poland boasts some pioneering environmental legislation, for example a law on water quality dating from 1922 was one of the first European examples of the precautionary principle. Environmental concern grew in the 1970s, mainly in the scientific community and through the involvement of scientists in incipient civil organizations. A major piece of legislation was the 1980 Environmental Protection and Development Act, and an official study of extreme ecological hazards was carried out in 1983.

The Act of 1980 was an advanced piece of comprehensive legislation even when compared to other European countries. It is an early example of the polluter-pays principle and user charges for resource consumption. Charges were designed for a wide range of emissions above legally permissible limits, and fines were established for breaching them. The standards set for legal limits were themselves very stringent (considerably tougher than those being applied in the USA at the time, for example). The Act also contained provisions for spatial development plans as well as environmental impact analysis, and was enforced by the National Environmental Protection Agency. It is not surprising that some Polish writers prior to 1989 (for example Starzewska, 1987) presented a very assuring picture of the country's environmental policy on the basis that it provided a set of legal controls at least equal to those found in advanced capitalist countries.

Interestingly, the Act addresses a situation which theoretically could not exist. The asserted superiority of planned socialist society over the anarchy of capitalism, embracing an economy based on the principle of production for society's good rather than exploitation for profit, must result in an environment which serves the best interests of society. Planning the whole production process would eliminate waste, optimize environmental transformation and guarantee the protection of future interests. In its commitment to praxis, socialism was held to be the only system able to correctly manage the production of the environment. This is the epistemological and political meta-narrative of socialist environmentalism. More specifically, the Act confronted one of the central tenets of classical Marxist economic philosophy. In charging for resource use, the Act values objects which, according to the labour theory of value, are freely available inputs into production. Value is created through the labour incorporated in their manufacture, thus the concept of scarcity pricing for environmental resources faced a considerable conceptual barrier only partially resolved by taking into account the vaguer Marxist notion of nature as the ultimate source of wealth.

Whatever the theoretical shortcomings underlying Polish policy, these were not the major factors contributing to the often dire environmental conditions experienced in the country (Thomas, 1999). Rather, the immediate causes are more often to be found in the political structure and functioning of the state. There are several aspects worth mentioning in this respect. First, the multi-level relation which existed between state institutions and the Communist Party meant that each level and branch of state institutions was supervised by a parallel party body. Key personnel occupying influential government posts were selected through the system of *nomenclature*. Thus, the party was able to ensure that its priorities were adopted at all government levels, the overriding priority being the maintenance of its position of dominance.

A second aspect concerns the relation between the ministries controlling the production industries and those concerned with spatial/territorial planning and environmental protection. Given the priority awarded to industrial production and the fact that the industrial ministries controlled the bulk of investment funds, including those dedicated to the creation of social infrastructure in the areas around industrial locations, the superior power of such organizations is understandable. This situation is an example of the commonly observed conflict between sectoral and spatial planning agencies. The strength of the industrial sector ministries was further reinforced by the ideological significance of industrial growth. Cole (1998) proposes that the post-war growth of the industrial economy under communism served as a mode of political legitimation in Poland, although Offe (1996) argues that this mode operated only in Czechoslovakia and East Germany and that the principal mode of legitimation in Poland was national identity.

Whatever the case, whenever territorial policy makers, obliged to consider the environmental impact of industrial activity on the environment, opposed the wishes of sectoral investors and operators, they almost always lost. In terms of the levels at which the charges and fines were imposed under the EPDA, the dominant influence of the sectoral operators ensured that these were set at so low a level as not to inhibit their operation. Rather than setting fines at a high enough level as to encourage the abatement of pollution emissions through improved production processes or the introduction of end-of-pipe technology, plant managers merely paid the fines as a production cost. Even if the fines and charges had been made at a significant level, the soft budgetary constraints under which the industries operated would have led to them being compensated from national allocations.

Finally, it should be observed that the law itself had a status different from that in Western legal theory, in that the state and the Communist Party represented the embodiment of the highest principles of the workers' state. As a consequence, judicial and statutory regulation assumed a secondary role and could be flexibly enforced whenever the situation required. In practical terms, the requirements of socialist development could be used to justify both the scarcity pricing of free natural resources and the systematic evasion of their effective implementation. In fairness, early attempts at natural resource use charges in the West experienced similar difficulties but without the politico-philosophic angst. In socialist countries it was ideologically more pressing to ensure concrete improvements in the lives of the workers – in employment, housing, health and education – than invest scarce resources in environmental management, especially when severe environmental problems could not exist, at least theoretically. Improving the material conditions of people's lives was also part of a political strategy to weaken social forces opposed to the workers' state. Ironically, these nominal concessions to environmental regulation outside the party system and the tolerance of environmental protest would be a major factor in the transition to a market economy.

Alternative voices: challenges to the party narrative

Whilst the Communist Party embodied the dominant ideology, institutionalized explanation and practical management of the relation with nature, it did not have a monopoly in this respect. Poland has some of the oldest nature protection organizations as well as an outstanding tradition in the natural sciences. Scientific concern about the environmental deficiencies of socialist development was important in revealing the deteriorating environmental conditions that the secretive ranks of the state and party apparatus were very reluctant to divulge. The incipient ecological groups which emerged in the 1970s sprang largely from their ranks. In particular, academics from Krakow, the 'intellectual capital' of Poland, were key figures in setting up the Polish Ecological Club in 1980. Emboldened by the rise of the Solidarity movement, significant victories were made on local issues involving industrial contamination. Environmental reforms then occupied a central place in Solidarity's famous 1981 demands.

However, as Wodz and Wodz (1998) have commented, environmental protest was as much a political act as an expression of ecological concern. The Communist Party implicitly recognized this. After the imposition of martial law and the suppression of Solidarity at the end of 1980, the state tolerated environmental groups (which continued to expand locally and organize nationally), sponsored and published its own high-level government reports, created an Environment Ministry and sought to contain environmental protest by creating the Social Movement for the Environment as an official umbrella organization. All this can be seen as an attempt to pre-empt protest and legitimate itself in the one area where political dissent was allowed. By the time of the 1989 Round Table talks called by the Communist Party to discuss the increasingly critical situation in Poland, environmentalists and environmental issues were strongly positioned.

Second, although the regionally-based and highly heterogeneous environmental movement achieved political status, it failed to capture the popular imagination (Ferry, 2002). The environmental movement was an intellectual as well as political one, and had little effect on popular social consciousness. The strategy of industrialization and the creation of an industrial proletariat had the effect that large sectors of society which directly benefited from these processes were little inclined to examine its environmental costs. On the other hand, large parts of the country remained traditional rural societies heavily influenced by the church and a religious conception of the world and nature.

Third, the rise of environmentalism has been associated with Polish nationalism. In general in the communist bloc countries, the environment was a source of anti-Soviet and anti-state protest which appealed to national independence. Tickle and Welsh (1998) argue that nationalist sentiment was much less in evidence in environmental protest in Poland than in other countries under Soviet tutelage, somewhat surprisingly so given the 'typical mythic construction of Polish self-identity as part of a continually oppressed and invaded nation'. Nevertheless, it has been argued that with national autonomy, and in the face of the harsh realities of the first transition years, the internationalization of environmental issues such as transfrontier pollution were constitutive and mutually supportive of nationalism. This could be equally applied to the adversely affected urban proletariat or in the countryside still governed by religious belief.

The point to be emphasized is that during the 1970s and 1980s the environment emerged as a reference for political dissent as much as widespread ecological concern. Tradition, nationalism, science and religion became reinvigorated as alternative meta-narratives to socialism and found political expression through the environmental problematic, in whatever of its many facets. As has been widely commented, once socialism was abandoned, the environment lost much of its momentum and political potency. In the post-communist period, in Poland as elsewhere in the former Soviet bloc, the emergence of democracy led to environmental leaders frequently moving into government positions and the fragmentation of an already heterogeneous movement. The environment as such ceased to be an exclusive outlet for opposition politics, and the processes of marketization, privatization and Westernization opened up many new opportunities for identity politics. What is more, newly acquired national autonomies sometimes resulted in a new hostility to the environmental movement, as opposition over issues such as power generation were held to be contrary to the national interest of economic growth. In the case of Poland, certainly new strains were opened between local government and the environmental movement (Kozakiewicz, 1996).

In general terms, the environment was part of a 'double rejective' (Holmes, 1997) of both Marxism–Leninism as a political philosophy and the domination of the Soviet Union over the Eastern bloc countries. However, the rapid collapse of the communist régimes left a policy gap in terms of an established, well-articulated vision of the new economic and social system to be established, a gap which the advanced Western states were eager to fill with their own highly ideological version of capitalism. The less dramatic unrolling of events in Poland compared to other countries in the region meant that whilst the environment lost some of its political prominence in the complex process and uneven development that ensued (Thomas,

1998), the environment also gained a new political significance. As Poland embraced the capitalist notion of sustainable development with enthusiasm, a whole new environmental protest movement was generated, now of an anti-capitalist rather than anti-socialist inspiration.

Institutions and the (de)mobilization of meaning in the transition period

Lodz may not have been one of the main centres of environmental protest (compared to say Krakow, Upper Silesia and Gdansk) but is now fairly typical of what is happening in Polish cities. With transition the environment lost its earlier political edge. The free conformation of political parties and unrestricted association meant the disappearance of the hitherto privileged position of the environment as a site of opposition. Throughout the 1990s, new associations and foundations flourished in Poland as a whole, and environmental organizations were, in numerical terms, relegated to the second division of NGOs, way behind those in areas such as education, health, social welfare and culture (Kurczewski and Kurczewska, 2001). This political repositioning of the environmental movement was part of the rebuilding of the institutional architecture in the post-communist period, and with it the authoritative sites for talking about and mobilizing meaning of the environment.

In this section we examine institutions in terms of all the formal organizations which influence urban development. In the case of Lodz it is clear that careful attention needs to be given to the different geographical and political scales. At the international level, transition meant not only the break-away from a crumbling Soviet empire but also a radical change of direction towards Western Europe and the European Union. European institutions can therefore be expected to have had a considerable influence on urban development policy and the environment. At the national level, transition to a market economy involved the wholesale reform of state institutions, the rise of the private sector and new forms of governance, and a general reshaping of the policy-setting framework. At the local level, urban governments were restructured and given greater autonomy in the moderately decentralized form of the new Polish state organization.

The shift towards the European Union and environmental internationalism

During the 1980s, the Polish state and Communist Party had made unique concessions to the environmental movement, which also played an important role during the 1989 Round Table talks that established democratic elections for the following year. This was essentially an internal affair, the particular Polish response to changes within the Soviet Union and the break-up of the communist bloc. What it did do was to unleash a strong desire for national autonomy and ultimate freedom from the restrictions of what in any case had been a non-monolithic Communist Party influence. Above all, the future was seen to lie, once again, in the West and the promises of capitalism.

The Round Table talks ensured that transition in Poland was a gradual and peaceful affair, and during the rigours of 'shock therapy' in the early 1990s the Communist Party shared power in the first period of democratically elected government. As for the environment, there was general consensus over the adoption of ecodevelopment as a strategic aim in the restructuring process. The notion of ecodevelopment differed from the more cultural, endogenous growth understandings of the term being developed in Latin America. In Poland its meaning was derived essentially from a reaction against the communist exploitation of nature and environmental deterioration under 'really existing' socialism. As a consequence, ecodevelopment was heavily inclined towards the protection of the natural environment over and above all other considerations. Just as this had been the overt basis for political protest under communism, so it became the central theme of post-communist development policy.

New environmental institutions, policy and instruments flourished in the early years. The Ministry for Environmental Protection, Natural Resources and Forestry was established in 1990, along with a national inspectorate and regional offices to control the implementation of environmental policy. The National Ecology Policy was adopted in 1991, prepared in the light of preparations for the Rio Summit and the idea of sustainable development. The abundance of environmental legislation passed in the following years was tilted towards unification with European Union standards, but not completely circumscribed by them. The country eagerly signed up to international conventions, regional cooperation programmes and pushed its environmental agenda and achievements on the international stage, extending its vision to Eastern and Central Europe and the developing world. Poland worked hard at using the environment to position itself both within Europe and on the wider international stage, and achievements in the environmental field were a card that Poland played geostrategically, as illustrated in the following claim by president Aleksander Kwasniewski:

> Poland, a country that in 1989 was the first to step on the way of transformation of political and economic systems, remains one of the global leaders in promoting sustainable development ... Poland's report on the 5-year long implementation of Agenda 21, covering all its 40 chapters, met with great appreciation during the 19th UN General Assembly Special Session five years afer the Rio Summit. Periodic economic and environmental performance reviews made by the OECD have shown stable and sustainable progress of our country's implementation of the Rio principles ... We wish our experiences, particularly in financing environmental protection and sustainable development, to be used on a much larger scale. These experiences have already been conveyed and successfully adapted by other Central and Eastern European countries. Our model of environmental funds could be easily implemented also in the developing world, particularly in those countries where appropriate legal and institutional structures exist.
>
> (Polish Ministry of the Environment, preface, 2002)

In effect, political legitimation no longer lay in the state demonstrating its sensitivity on environmental issues to local organizations but, rather, its commitment to external institutions such as the European Union, multilateral bodies and international NGOs. As a result, foreign institutions increasingly influenced national environmental policy. Environmental NGOs from countries such as the USA, Canada, Holland, Germany and Sweden moved into Poland with enthusiasm, cooperation funds from Europe became more important, and legislation and regulation of the environment was explicitly aimed at harmony with the European Union after 1994. The economic benefits were not that significant, since 95 per cent of investment in the environment continued to come from domestic sources (Polish Ministry of the Environment, 2002). Foreign financial aid was mainly in the field of institution building and technical training, which meant not only greater expertise in environmental management but also the discursive take-over of its social and political meaning.

A good example of the new official environmentalism is Lodz' participation in the joint World Health Organization/European Healthy Cities Programme. In the early years of transition, several cities in Poland joined international networks such as the International Council for Local Environmental Initiatives (for example, Gdansk) or signed the Aarlborg Charter of European Cities and Towns entitled 'Towards Sustainability'. Lodz was the first Polish city to take part in the Healthy Cities initiative and plays a leading role in the network of 30 or so cities now members of the Polish Healthy Cities Association. The European Union sets the agenda and performance standards, which are then used by the urban authorities to promote the city's image and legitimate local government. This is not to dismiss the value of the programme itself in Lodz, which has a dynamic organization with some important tangible achievements. However, from our theoretical perspective it is clear that the programme voices the European sustainability agenda from a health angle, and with it all the assumptions behind a market-led policy of ecological modernization. In this sense, environmental health has a barely-concealed political objective of preparing citizens for a capitalist society which makes increasing demands on the personal performance of individuals. The presentation by Krysztof Panas, Deputy Mayor of Lodz, in a recent report is illustrative in this sense:

> In accordance with the WHO definition 'health is a state of complete physical, mental and social well-being and not merely the absence of disease'. Health is also the ability to perform social functions, adapt to environmental changes and cope with these changes. Health is one of the basic human rights. It is:
>
> - a value, thanks to which people can realise their aspirations, change the environment and the conditions of life.
> - a resource for the community, guaranteeing its social and economic development.
> - a means to achieve a better quality of life.
>
> Good health of the community is the foundation for sustainable economic development, it is at the same time a way to measure social cohesion and balance.

> . . . It is my pleasure to inform you that at present it is the only document of this kind in Poland and one of the first in Europe. It promotes Lodz and creates the positive image of the city.
>
> (Lodz City Office, 2001)

Environmental programmes such as this bring with them international authority whose influence expands over extensive areas of local government management practice. At the same time, local environmental organizations are de-authorized and marginalized. In Lodz, the fluctuating fortunes of the Centre for Environmental Activities is a good case in point. In the early 1990s, the city authority let to this organization, part of the Polish Green Network, two premises on the main Piotrkowska Street, one a campaign office and the other a cultural centre. As the focus of legitimation shifted to the European Union, local organizations lost their political significance and, in this case, their accommodation. By the late 1990s the Centre for Environmental Activities had become so desperate with the delays, obstacles and rent increases imposed by the city authorities that it was forced out. The premises were given over to high-rent commercial uses, and what remains of the Centre now operates from a privately-rented, single-room office on the third floor of a building located on the city centre periphery. This is a symbolically powerful indication of the decline in the political significance of local environmental organizations in general, and the increasing mediation of relations through European institutions, whether national governments, the European Commission or NGOs.

When Lodz adopted its own Local Agenda 21 in the document *Ecological Policy for the City of Lodz* (Lodz City Office, 1997), this was firmly contextualized in the urban sustainability agenda of the time, with extensive reference made to the Rio de Janeiro World Summit, the Congress of Local and Regional Authorities at the Council of Europe and the Aarlborg Charter arising from the first conference of the European Sustainable Cities and Towns Campaign held in 1994, to be implemented within the V Environmental Programme of the European Union. The central proposition that 'environmental protection cannot be treated as in conflict with the interests of the national economy but rather as a question of proper management' reflected both European policy and general thinking amongst the development planners of transition in Lodz (Markowski and Rouba, 2000). A deteriorated environment, it was held, had a negative effect on the city's image and was not conducive to inward investment. Economic restructuring would facilitate environmental improvement which in turn would stimulate further economic growth and a better quality of life in the city – the virtuous circle of sustainable development. In its recommendations for the future, the spatial dimensions of the policy are based on a plan prepared in 1977, dosed with the new economic agenda of heritage marketing, more roads and motorways, special economic zones and ecotourism.

The ecological policy document acknowledged a lack of participation in its formulation, arguing that time constraints imposed by the city council made this impossible to carry out. One or two environmental experts from local universities were involved, along with the conservative League of Nature Conservation and the Polish Society of Town Planners. However, communes and NGOs were substantially ignored, but it was argued in defence that the document should be understood as a starting point for a wide and vigorous extension of environmental issues into all aspects of life in the city and the future participation of all organizations that actively promote civic activity. However, such neglect of NGOs would have been unthinkable at the beginning of transition.

Meanwhile, the informal coalition of environmental organizations in Lodz had been breaking up. Once unified against the socialist system, it splintered in the face of new options, opportunities and interests created under transition. There was, however, an ever-present mediation of external institutions and the influence of Western Europe. The state gradually withdrew its financial support to environmental organizations, which became increasingly dependent on funding from foreign organizations or contracts for providing direct services in fields such as environmental education. Marginalized by local government and supra-national programmes, the environmental movement retreated from politics into the fields of education, ethics and lifestyle issues or, alternatively, assumed a more radical anticapitalist and anti-globalization stance. The underpaid scientific establishment, so influential in providing the information and analytic base for early environmental protest, became seduced by the opportunities of European integration and consultancy work as expert advisors to the new local authority. Above all, the rules of the game had been set outside Poland in the policies, directives and regulations of the European Union. This was now the site of ultimate authority on environmental policy and the institutions of local government and all those who cooperated with it were now bound its rules.

This did not mean that the local authority itself played strictly by those rules. The free flow of information proposed in the city's ecological policy failed to fully materialize, and rather than becoming a key area of public debate on the future of the city, environmental issues became locked into a grid of suspicion. The general feeling was that environmental interests became all too easily sacrificed in the market processes of urban development, with politicians and city officials frequently involved in unclear liaisons with private developers. Environmental groups complain of obstacles in access to information and the begrudging attitude of the city authorities to the support of local initiatives. Since 1997 no new comprehensive policy document has been produced, despite its original provisional nature, and environmental awareness amongst ordinary citizens remains low.

In short, the move towards the European Union ensured a paradigm change in the conceptualization of the environment and its place in the urban development process. Once the domain of scientific socialism and party dogma, throughout the 1990s the environment gradually became re-situated in the framework of Western sustainability discourse, private enterprise and individual responsibility. European institutions assumed the authoritative voice and regulatory prerogative on environmental matters, with important implications for the institutional position and power relations of the city authorities and local organizations. Institutional reform at the city level facilitated this process.

Decentralization and urban environmental management

Transition necessarily involved the restructuring of state organization and the institutions of public administration. The highly centralized system of government dominated by the Communist Party was replaced by democratic political institutions and the decentralization of government functions. Our interest here focuses on the urban level and the redistribution of power and responsibility with regard to urban environmental management. However, initially a brief introduction to local government reorganization over the period is useful.

Until 1989, party officials and production managers were largely responsible for local development policy and investment in urban infrastructure. The Local Government Act of 1990 saw the dissolution of the old-style 'people's councils', which at the local level (communes) were replaced by democratically elected local authorities with considerable autonomy in running their affairs. At the provincial level, the voivodeships continued as non-elected regional expressions of central government, appointed by the prime minister. The local government system was further developed in 1999 when the 49 voivodeships were reduced to 16 and a three-tier model of territorial organization was introduced. The voivodeships or provinces were strengthened by becoming elected bodies with increased functions, *poviats* or municipal authorities were created in the larger towns and cities, and communes continued as the lower level of government. Local government reform has been seen as both political and economic: a consolidation of democratic local government and open and efficient public administration (Czyz, 2002), a response to the global economy increasingly based on urban regions (Klasik, 2002; Markowski, 2002), and a requirement for future access to European Union structural funds (Pyszkowski, 1998) and the implementation of new labour market policies (Churski, 2002).

In political terms it has been argued that the new local authorities were a stabilizing element during structural change and efficient bodies in the area of public spending. Less constrained by the old Communist Party apparatus, they were also effective in facilitating the privatization of state-owned enterprises and accommodating local society to the rigours of a market economy, at the same time distanced from the power struggles between the newly constituted political parties at the national level (Szul and Mync, 1997). However, political representation was much slower to take form at the local level (Wodz, 2002). In the first local elections of 1990, ad hoc Civic Committees were created to make up local electoral lists. The Civic Committees were typically led by local social and political figures backed by circles of earlier opposition groups. This meant that both national parties and regional associations initially had a limited impact on local elections. The Civic Committees facilitated the emergence of charismatic local leaders, the participation of special interest groups and the insertion of the private sector into local policy and urban management. However, with the passage of time the political parties re-established control and in Lodz local council membership is currently dependent on first getting onto party lists.

The new local authorities were also given important environmental responsibilities. It has been argued that initially they were diligent in assuming these functions and many large cities in Poland eagerly joined international urban environmental networks. However, in practice there were several difficulties in the early stages. National ecological policy was highly centralized, especially in two important areas: funding and regulatory controls. In terms of funding, the National Environmental Fund received income from environmental charges and fines, debt-for-nature schemes and central budget allocations, which were then redistributed to the local authorities via the voivodeships. The voivodeships or provinces, non-elected bodies until 1999, also exercised regulatory powers on environmental protection. The result was a difficult relationship between the voivodeships and the local authorities (communes), including disputes over land use and urban development decisions. An overburdened legislative agenda on all aspects of reform meant that these problems of environmental management persisted throughout the 1990s.

A second major problem area concerned the functioning of the communes or local authorities themselves, where it was not easy to establish an efficient local public administration. New-style charismatic leaders had to work with old-style local bureaucrats and a lack of expertise in many areas of public administration. Although many civic associations and NGOs arose, generally speaking urban populations maintained their passive attitude towards local government and distrust of authority in general. As a consequence, the participatory pretensions of urban environmental management encountered difficulties on all fronts, as evidenced in the formulation of the ecological policy for Lodz. A further point worth making concerns the instability of institutions, law and policy in the period of transition. This created a climate of uncertainty and lack of transparency in the administration of public affairs, encouraged public apathy and was propitious for shady or corrupt practices in local government. In short, the *realpolitik* of urban management was a far cry from the formal pretensions of the legislative framework, as we will see in the following section.

After the initial successes of environmental improvement, in large part due to the collapse and restructuring of the Lodz economy, public institutions for urban environmental management had to face up to the increasingly evident shortcomings of new local environmental policy at an urban level. Furthermore, from the ecological policy of 1997 onwards it was increasingly acknowledged that economic restructuring was creating new challenges for the sustainable development of the city outside the traditional sphere of natural resource protection. The rise of structural unemployment, poverty, social pathologies and consumerism threw the social dimension of the city's sustainability strategy into question, the threats to environmentally dirty industrial sectors provoked a backlash reaction from the working class and growing social differentiation saw the rise of a self-interested and reactive 'nimbyism'.

Local spatial planning and politics

As observed earlier, spatial planning during the communist period was largely a secondary concern, a consequence of the economic development planning carried out by party officials and plant managers. The Lodz voivodeship's spatial plan dating from 1977 was a regional development plan for the distribution of industry, population, infrastructure and services, not dissimilar to UK strategic plans of that time. The plan contained an environmental component which still informs current city planning in Lodz: a regional system of protected green open spaces penetrating the urban centre, the control and relocation of heavily polluting industries, better public transport, central area improvement, green corridors in river valleys to aerate the city, and a sewerage treatment plant.

The Local Government Act of 1990 established spatial planning as a primary local authority function. Spatial planning in the Polish context is a detailed form of land use and physical development planning, and in Lodz such a plan came into force in January of 1994 with a legally-binding status and a ten-year life span. It was formulated under the dual premises of a market economy and sustainable development which characterized national policy. The first one required that spatial development, rather than simply reflect economic planning, should actively promote it by establishing conditions favourable to private initiative and investment in the restructuring process. The second one constituted limits to private initiative and a set of basic functions for the local authority. The physical dimension of the new plan proposed the containment of urban sprawl, mixed-use multi-functional areas, small-scale private housing development, strict industrial zoning, the development of regional functions, city centre revitalization and the improvement of the city's image.

In the context of transition, the plan was both excessively rigid and overly flexible. The rigidity of the plan concerned its detailed land use designation which failed to appreciate fully the demands and dynamics of future urban development. The flexibility concerned the ease with which the city authorities could amend the plan, carried out on an *ad hoc* basis by the city mayors, generally in relation to negotiations on urban projects with private developers. The plan expired at the end of 2003 and the city authorities were not optimistic about being able to revise and update it in accordance with the demanding requirements of new national guidelines, opting instead for a policy statement and land use proposal which will, however, lack the full weight of legal authority. This would only add to the general uncertainty over an extended transition period concerning what is permitted or not, and what is legal or not. According to some commentators, a drift from 'total planning to total flexibility' means that Lodz will be unable to prevent residential growth into protected areas and control urban sprawl, despite the good intentions of the new plan.

Undoubtedly, the planning process in Lodz is increasingly oriented towards promoting private sector economic growth, as evidenced in a recent spatial economy study produced in 2002 which will exert a fundamental influence on future development. However, partnership arrangements with the private sector have been crude, un-institutionalized and somewhat speculative in nature. Past experience has shown that partnerships tend to be carried out on a personal basis by political leaders and that there is a poor level of coordination with formal initiatives such as the Special Economic Zones set up under national government supervision. As a result, environmental management tends to be subordinate to economic development, being seen as a condition for economic growth and subject to the precariousness of the regulatory system at both the environmental and land use planning levels.

It is perhaps this context which explains what appears to be a unique phenomenon in Lodz: the commodification of environmental protest through the legal system. With the political and administrative systems failing to ensure environmental guarantees, it is the legal system which has been most effectively used by the more independent environmental groups. During the late 1990s it became a common strategy for environmental organizations to object to urban development decisions through the courts, which involved long and costly delays for the developer. The outcome was either the eventual strict compliance with environmental standards and planning policy (for large-scale projects) or the illegal development of land outside the planning system (facilitated by uncertainty with regard to the legal status of certain types of land use regulation). Projects concerning large-scale capital investments were the most vulnerable to legal challenges and lengthy delays. Environmental groups capitalized on this situation not only to ensure compliance with environmental controls but also to challenge the very nature of the new trends in urban development. A celebrated case in Lodz concerns the Obywatel environmental NGO which, in 2002, negotiated an out-of-court settlement of around US$50,000 for withdrawing a legal challenge to a shopping development. Obywatel was quite open about this in the media, arguing that this was a legitimate way to secure funding for its activities and pledging open accounting on the use of this money. In the event, the development project, and most of the financial settlement dependent on it, fell through.

The significance of this case is considerable in urban terms, in the sense that it illustrates a general marketization of the environment. Environmental regulations themselves are sufficiently strong to ensure that they have a considerable impact on the economics of urban development for all parties involved in the process: for the local authority it is an important source of income, for private developers an important factor in land rents, for communes a bargaining chip in the sale of state-owned property, for politicians a source of illegal income through corrupt practices, for NGOs an area of political pressure if not financial resources. Once a free good, widely abused under socialism, the environment is now a scarce commodity extensively maltreated through the politics of the market.

This is not to imply that the environment has lost its importance in the development of Lodz. In fact, the commodification of the environment could only occur through the environment being actively constituted as a scarce good subject to strict regulatory mechanisms, a process not exclusive to Lodz but which is thrown into sharp relief in a transition context. However, it is a distorting factor in the aim for sustainable urban development in that the scientific rationale of ecology is replaced by the principle of competition and the uneven distribution of political power over space, the consequences of which we will examine in the following section.

Spatial form: getting into (the environmental) line?

In general terms it can be argued that on the eve of transition Lodz enjoyed excellent conditions for moving towards a sustainable model of development as understood in the light of contemporary urban agendas. It was relatively compact, with ample open space provision accounting for around 15 per cent of the urban area. It had an extensive public transport system of trams using electric power, car ownership and car pollution levels were low and congestion non-existent. Combined heat and power generation from the city's power stations covered approximately 60 per cent of all households in the city. In social terms it fulfilled the aims of equality and social justice as understood in socialist terms. Deficiencies mainly concerned the state of natural resource systems. Air and water pollution were high, much of the central area and the older housing stock were in poor condition, and public health was unfavourable compared to European standards. Public participation in local government and environmental management was low, though theoretically unnecessary under the earlier socialist system.

In principle, it might also be argued that transition to capitalism and all that this implied should have consolidated these sustainability advantages and helped to overcome the outstanding problems. Transition opened up the possibility of inward investment and the technological modernization of industry and, with entry to the European Union on the horizon, the obligation to implement strict regulatory requirements with regard to natural resource protection. It also involved democratic (elected) local government, a strengthened civil society and greater participation. Certainly, these were the assumptions behind much academic thinking and official policy in the early 1990s. What really happened in urban terms? In this section we examine the trends in urbanization and spatial form in the transition period and analyse whether Lodz effectively moved towards a more sustainable form of urban development.

The new spatial economy

The shock therapy of the early years of transition was traumatic. Nationally, inflation rocketed to 2500 per cent at the end of 1989, unemployment soared from 0 to 16 per cent by the end of 1993, average real incomes fell by a quarter between 1989 and 1990 and carried on falling until 1994 (Markowski and Marszal, 1999). In Lodz, the textile-based economy was especially vulnerable, and unemployment in this sector reached 20 per cent by the end of 1993. Job losses were severe in the manufacturing sector as a whole. Fixed prices and state subsidies were dismantled, industries were uncompetitive in the face of inflowing goods and services, profitability plummeted, and labour cutbacks and firm closures were inevitable. In 1994, all of the communes in the Lodz voivodeship were included in the list of communes with exceptionally high structural unemployment (Markowski and Marszal, 1999). An obvious corollary was a sharp increase in poverty as job losses exceeded the capacity of the social security system. State budgets in general were slashed and public investment curtailed. Furthermore, psychological factors played an important role. The guarantees and stability of the socialist system were disappearing, more often than not precisely for those people who most relied upon it: the industrial proletariat.

The spatial effects of those early years included the large-scale closure of industrial plant and the emergence of a new element on the urban landscape: disused factories and derelict industrial sites. Public spending was cut back due to reduced local contributions of state-owned enterprises or dwindling central government funds, and housing maintenance continued to deteriorate. The introduction of commercial pricing and the metering of public services saw considerable increases in the consumer costs of water, electricity and transport. Not only the economic outlook but also the urban and social landscape of Lodz was bleak indeed.

Plant closures, privatization and labour cutbacks were to some extent compensated by the creation of new firms. In the early years of transition, state-owned firms tried to ameliorate employment effects by reducing shifts and overtime hours and by introducing involuntary unpaid leave. After 1993, however, the imposition of market principles in state-owned firms as well as privatization accentuated the fall in industrial employment (Markowski and Marszal, 1999). One of the main reactions was the creation of new, more flexible small businesses and a major transfer of employment to the private sector. By 1996, 65 per cent of industrial employment was in private firms, reaching around 85 per cent in the construction and retail sectors. The textile and clothing industries were able to increase their participation in European markets and recover some ground lost in the former Comecon countries. A corollary of this was the establishment of large 'bazaars' or open markets on the edge of the city, largely wholesale markets for direct producers with thousands of stalls and attracting clients from all over Poland.

Foreign investment was initially sluggish in Lodz. By the late 1990s some foreign companies such as ABB, Coca-Cola, Pepsi, Shell, Legler and Statoil had set up in the city. To encourage foreign investment, the Polish government created Special Economic Zones for the most deprived regions such as Lodz. These are mixed public companies financed through central government and run by a management board of national, regional and local representatives. Their main function is site provision for national and foreign companies, on publicly-owned land. There are ten such zones in the Lodz region, both greenfield and brownfield, which had created some 4000 new jobs by 2002.

The attraction of new firms created spatial dilemmas. Most new companies required medium to large well-serviced sites. Unused buildings or derelict land was abundant in the old industrial districts, but did not always meet the location requirements of new companies. The Special Economic Zones helped assemble and service some brownfield sites and could offer tax and other incentives. However, the city was under great pressure to succumb to the demands for aesthetically attractive sites with good road communications on green land both within and on the outskirts of the city. This demand was especially strong in the retail sector, which has been largely taken over by foreign companies. Lodz now has some 20 hypermakets (Tesco, Carrefour, Macro, Geant, Jumbo, Leclerc, etc.) spread around the city's intermediate ring road or on the major roads of the city periphery.

The transition to a market economy was clearly not just a matter of privatization. It required and brought with it a new structure of spatial organization based on private capital investment and competitive location criteria, which most obviously affected the industrial and retailing sectors but also had significant implications for an emerging housing market and the social consumption of space.

New socio-spatial divisions

Transition led to population decline. The city's population had grown steadily in the socialist period, especially in the 1970s. However, after 1990 population began to decrease through both out-migration and a fall in marriages and the birth rate. The population fell steadily from a peak of around 850,000 in 1989 to 786,000 in 2001. Economic restructuring was a major factor, as job losses drove people to seek opportunities in less affected areas or return to family farms in the countryside.

Transition also deepened regional disparities and intensified social polarization. In place of the economic inefficiencies of the planned regional distribution of growth, market forces led to a concentration of new economic activity in the most profitable (competitive) locations. Market-determined wage structures, aggravated by business cycles, led to growing income differentials both within and between regions. In Poland as a whole, the distribution of employment in the private sector passed from 12 per cent in 1989 to 73 per cent in 1999, excluding the black economy which accounts for an estimated 20 per cent of national income (Weltrowska, 2002). The introduction of a market economy led to the emergence of a diverse group of entrepreneurs, many small businessmen of peasant origin, working people and the intelligentsia, as well as the old and new (post-Solidarity) nomenclature in a privileged position in the privatization process. These power elites had been consistently successful in accumulating capital, with the skills and experience necessary to take advantage of the new and legally ambiguous situations. Income differentials that varied by about 30 per cent above and below the national average in 1989, had by 1993 increased to 260 per cent above and 40 per cent below. These trends have accentuated enormously since then, whilst structural unemployment has remained at about 10 per cent, of which only 24 per cent received unemployment support in 1999, compared to 71 per cent in 1991 (Churski, 2002).

In Lodz, industrial employment in the voivodeship fell by over a quarter between 1990 and 1992, with industry accounting for 41 per cent of total employment in 1990, 35 per cent in 1996 and 29 per cent in 2001 (Markowski and Marszal, 1999; Lodz City Office, 2002). Labour market policies were developed allowing some 60,000 people in the region to participate in job training, public works and loans schemes. However, job opportunities were scarce and wages in Lodz are only 60 per cent of those in Warsaw and approximately 15 per cent below the national average. A blind eye was therefore turned to an increasingly significant black and grey economy, especially in trade/bazaars and small business activity, although there are no reliable statistics available to measure its impact on incomes in the city At the other extreme, a new business elite emerged, often associated with foreign investment, along with a high-level service and consultancy sector, and a revitalized nomenclature to oil the wheels of public-private partnership. In general, the explosion of top-end incomes is in plentiful evidence in the luxury cars, Western-style single-dwelling housing developments and expensive shops and restaurants which have mushroomed in the city.

In terms of social structure, this meant the emergence of a new social elite around a privatized economy, a struggling working class, a vibrant informal economy of small businesses and, left-behind, the elderly, long-term unemployed and unqualified young people – the least-prepared or worst-situated to adapt to transition. Perhaps inevitably, alcohol and drug abuse rose along with poverty and social tensions, currently the cause of considerable concern. The city no longer considers itself a safe place in the face of a systematic increase in registered crimes (Lodz City Office, 2001), although serious crime levels are still relatively low.

The housing market and spatial consumption

The emergence of a housing market is one of the clearest indications of social differ-entiation. In the socialist period it was difficult to move from one housing estate to another, the differences between them were not that significant anyway and people tended to develop emotional ties to the places where they lived (Liszewski *et al.*, 1995). Manual workers and professionals shared the same housing develop-ments in basically equal conditions. However, since 1989 unemployment and income inequality have strained good neighbour relations, and those with higher incomes tend to seek better accommodation. Although new forms of association were devised to secure housing maintenance and improvement on the older estates, the emerging ideal is a detached house with garden, despite the fact that it is beyond the means of all but a small social group.

The development of a housing and property market was a fundamental dimen-sion of transition, since the private ownership of land is an important area of finance capital circulation (through savings, loans and the real estate sector) as well as an ide-ological tenet of capitalism. Initially, the housing market was hampered by unclear property rights, the lack of institutional and financial structures, an undeveloped building industry and uncertain land use and legal provisions. Together these factors led the slow take-off of the housing market and a preponderant role of the local authorities through the control of state-owned land. They tended to facilitate a spec-ulative property market and unplanned city expansion towards underdeveloped greenfield sites. By the beginning of the new millennium these obstacles had been largely overcome, to such an extent that (ungated) low-rise private housing estates and single-plot developments now dominate much of the urban periphery.

The new social elite brought with it a wider change in the cultural consumption of space. The new imaginary of individual consumption under capitalism, culturally defined and substantially constructed by transnational corporations and privatized television networks, saw the rapid development after the mid-1990s of hypermarkets, shopping centres and American-style strips. This new form of spatial organization depended, of course, on the private car, whose numbers increased vertiginously from 130,000 in 1990 to around 280,000 in 2001. Spatial specialization and expanded consumption were not confined to the city itself, as conspicuous consumption under transition – and the opening out towards Western Europe that this implied – also included a renewal of the ideal of a second home in the countryside, foreign travel and holidays abroad. The ultimate liberation from restrictive socialist space may be seen in the mobile phone, a potent symbol of contemporary capitalist lifestyle which consumes a disproportionate amount of salaried employees' income.

City image and place promotion

After the initial upheavals of transition, city promotion came on to the urban agenda, and the image of Lodz as a dirty industrial centre was a problem. Markowski and Marszal (1999) argue that Lodz enjoyed several advantages in the transition to a market economy. Private enterprise had never completely disappeared under communist rule, it had a relatively skilled labour force and strong local institutions which were able to help respond to the challenges of market forces. However, it suffered from poor road and air communications, and great expectations are currently placed on the development of a national motorway system which places Lodz at the major central junction, thereby constituting an important location factor for industry and the distributive trades. Financial constraints have prohibited its development so far.

Nevertheless, Lodz attempted to develop its tertiary sector – shopping, trade, finance, insurance, culture and tourism – all heavily dependent on an attractive and marketable city image. This meant, above all, revitalization of the city centre. The central area comprises the large nineteenth-century sector which was largely abandoned during the communist period. It has considerable architectural and urban quality of a non-monumental kind, but is affected by physical decay and complex property and occupation rights and is inhabited by the poorest sectors of urban society. Apart from the main Piotrkowski Street, its grid-like urban landscape is still blackened by earlier industrial pollution and dereliction.

From the beginning of the 1990s, Piotrkowska Street was identified as a key element for improving the city's image. Traditionally the centre of institutional and cultural life in Lodz, most of the major administrative and cultural activities are still located along or off this axis (Wolaniuk, 1997). Building restoration and pedestrianization were accompanied by new shopping and leisure activities. What was

6.3 The main Piotrkowska Street, which runs the length of the historic centre, is now bright, largely commercial and fitted out for tourists.

6.4 New shopping malls encapsulate all the consumer attractions and sustainability contradications of Westernized Lodz.

until very recently a sedate high street is now a colourful space with shops, offices, bars and restaurants spreading into the side-alleys and courtyards. Bicycle taxis carry people along its four kilometre length dotted with outdoor bars and restaurants. Urban regeneration is, however, strictly confined to this artery, and one or two other pieces of exceptional architectural and urban heritage, such as the Poznanski Palace and factory complex now being renovated as a shopping and cultural centre. The city is working on the revitalization of a wider central area, although the invest-ment requirements are huge and there seems little likelihood of attracting upmarket housing, at least in the immediate future.

Along with urban heritage, cultural activities are a vital part of contemporary image promotion. Lodz currently places great emphasis on its cultural history and both traditional and modern festivals of all kinds, including the international figures associated with them, such as Arthur Rubenstein and Roman Polanski, an honorary citizen. Of especial importance is the Four Cultures Festival, celebrating the historic links between Polish, German, Jewish and Russian peoples. At the other extreme, Lodz has something of a reputation for low-budget experimental and alternative arts which rivals the Warsaw scene. In the coldly calculating strategies of the culture industry, the famous Jewish cemetery (the first ghetto under Nazism was established in Lodz) is suggested as important for 'emotional tourism'.

The environment in place promotion

Contrary to what might be expected from its formal political importance, the environment tends to be a very secondary issue in the current promotional litera-ture. The glossy edition *Leave Your Heart in Lodz* (Sagalara, 2002) certainly pushes history, heritage, culture and business much more strongly. It does mention recent environmental achievements such as the improved technological monitoring of air pollution, the initiation of a waste water treatment plant and the introduction of waste materials recycling and, in the voivodeship, Poland's biggest plastics recycling plant, over the period 1992–95, and later improved technology in boiler houses and power plants, new Mercedes buses complying with European environmental norms, improvement to tram tracks, three new hospital waste incinerators, and 'intensive actions aimed at solving the problem of the new municipal landfill site'. It also emphasizes the recognitions received, such as the tram company European Medal for Services in 2001, awarded by the Office of the Committee for European Integration and Business Centre Club, and its 'healthy city' award from the Lodz authorities. For its part, in the *Lodz Business Guide* (2003) the state-owned Heat and Power Plant Group celebrates its coverage, efficiency, new technology and ISO-9001 status, and concludes with the statement that the Group ZEC-SA 'is just on the eve of privatization – shares should be made available to a strategic investor in the second half of 2003'.

General promotional literature is aimed at a wide audience, but neither there nor in the business sector is much attention paid to the ecological environment. Industrial and business investors are primarily interested in strategic location, labour markets and the effective cooperation of the local authority; for manufacturing, the image of the city is of secondary importance. As elsewhere, the place promotion strategy has been questioned in Lodz, and its effectiveness in terms of local economic development and social equality issues, such as who actually benefits from flagship projects such as Piotrokowska Street. Certainly the services and prices there exceed the consumption capacity of the average Lodz citizen. It also happens to be the most dangerous part of the city as far as petty crime is concerned. On the other hand, urban imagery can be important for tourism, but this is an underdeveloped sector in Lodz. Young and Kaczamarek (1999) have analysed some of the eccentricities (and admirable frankness) of early place promotion material in the city (for example, 'It is a pity that the two largest rivers in the province are practically conduits for municipal sewerage' – Lodz Tourist Chamber, 1997), and argued its importance in terms of changing the *local* population's attitude to their own place to assist in economic, social and political transition.

What limited attention the environment receives seems to concentrate on technological achievement and Europeanization. This supports the thesis that the environment is principally a question of image and legitimation, both from outside (Europe and transnational corporations) and from inside (citizens and their mental image of the city) in an effort to demonstrate the modernization and attractiveness of Lodz. This is all a long way from any serious consideration of sustainable urban development and the new conflicts which have arisen through the transition to capitalism.

Urban environmental conditions in the new millennium

In this section we look briefly at the objective state of the environment in Lodz entering the twenty-first century. The discussion of the environment in the city has often been framed in terms of the superior environmental performance of private enterprise and market mechanisms over the socialist centrally-planned economy. In this area, as in much else when discussing the contemporary history of the city, the achievements of the socialist period tend to be ignored or forgotten. We will attempt to re-establish a more balanced perspective as a prerequisite for assessing the current sustainability of Lodz.

In the case of air quality, for example, this certainly improved in the early years of transition. However, significant improvements had also been achieved under socialism despite the general under-investment in urban infrastructure. There were two main sources of air pollution in the city: power and heat generation and industry. In the years 1945–60 housing estates were equipped with local boiler houses with no emission controls, but in 1957 a centralized system of domestic heating and hot water was initiated on the basis of surplus heat from power plants. Chimneys were also built higher to encourage dispersion, dust removal was improved and plants were made more efficient in terms of the low-grade coal combustion. Local boilers were gradually shut down as the distribution system was built up and all new developments were connected to the city distribution system, which by 1990 had an extension of 600 km. Nevertheless, total power plant emissions did not begin to decrease until after 1980, and only then somewhat marginally. This had much to do with the poor maintenance levels of filters and central government directives requiring the use of the lowest quality coal which exceeded technological cleaning capacity. Much more significant reductions were achieved in terms of emissions from industrial and residential boilers from the 1970s onwards. However, in 1990 individual stoves and fires using coal and wood still accounted for about 30 per cent of heat demand, contributing significantly to dust and carbon dioxide emissions. Industrial pollution was also reduced in the socialist period by some technological improvement in production processes and, after the 1966 law on air pollution, the concentration of industry in industrial zones, the removal of some heavily contaminating industries from housing areas and the introduction of end-of-pipe treatment of waste. The general picture was one of good basic systems lacking technological development and effective implementation.

In the early years of transition, severe economic recession contributed to a more accentuated reduction in resource consumption and pollution levels. The collapse of the textile industry caused drastic reductions in water and energy demand, which later fell even further by reduced domestic consumption through metering and higher charges. The dramatic reduction in water consumption meant that there is now over-capacity in the water treatment plant and a rising water table, which some commentators fear will lead to new problems as underground water supplies come into contact with dangerously polluted soil. New industrial enterprises were able to introduce cleaner technology, stricter pollution control was enforced and the state-owned heat and power plants were significantly improved. The overall result was that during the 1990s, even after the recovery of industrial production after 1993, air quality continued to improve, especially in terms of dust, sulphur dioxide and carbon dioxide.

Since 1991, mean annual dustfall levels have been within accepted limits, with the city centre having benefited especially, but concentrations of suspended particulates still exceeded permitted levels on numerous occasions up to 1995. The city centre remains the most polluted area. Sulphur dioxide and nitrogen dioxide levels also tended to decrease, although in the latter case reduced emissions from power generation were offset by increasing vehicle emissions. The effects of growing car ownership were also evident in higher levels of pollutants such as lead, ozone, formaldehyde, and benzo-a-pyrene and tar substances over the 1990–95 period (Lodz City Office, 1997).

On the other hand, some old environmental issues remain unsolved and new ones have arisen. The sewerage treatment plant and solid waste disposal continues to be a serious problem, along with the coverage and efficiency of waste separation and recycling. More worrying in the long term is the exploding model of spatial organization. The marketization of urban development has resulted in the eating away of the city's green wedges and significant expansion onto peripheral greenfield sites, despite the 7 per cent fall in population since 1990. Investment in road infrastructure has vastly exceeded that in public transport and environmental groups accuse the city of purely cosmetic improvement of the tram system. Car pollution, noise and congestion are growing problems, exacerbated by regional traffic and heavy goods vehicles having to pass through the city. Questionnaire surveys undertaken in the mid-1990s revealed noise emanating from domestic sources and old lift and sanitary installations to be major causes of nuisance, exacerbated by the poor sound insulation of much housing stock. Despite improbable official claims that Polish citizens unconsciously regulate their own lives in an environmentally sensitive way (Polish Ministry of the Environment, 2002), it is generally felt that environmental awareness in Lodz is low and that environmental NGOs now have little impact.

The condition of the environment is therefore being increasingly framed in terms of quality of life issues as opposed to the earlier, more ecological, approach. It is precisely this aspect that the more radical environmental groups attack: the idiotization of consumer culture, the fetishistic love-affair with the car, the invasion of foreign-owned supermarkets, the banality of privatized television, social inequalities and the shift, concentration and corruption of power in the city.

Another point worth emphasizing is the effectiveness of market mechanisms. Certainly, under democratic local government these have been more strictly enforced. However, as we saw earlier they were introduced in the socialist period and new signs of weakness are appearing under market pressures, whilst other sources of pollution seem to be outside the technical scope or political will of government, including the problems associated with car pollution and urban expansion. Other environmental benefits resulting from reduced natural resource consumption have been financed by the general population through heavy increases in the cost of public services.

Conclusions

The analysis of Lodz has revealed an urban environment subject to many changes over a period of system change, involving its conceptual construction, political management, social evaluation and physical condition. In this final section we will attempt to draw together some of the main features and address the question as to whether, in any meaningful sense of the term, the city is more 'sustainable' than before.

One of the most striking features of the Lodz experience is epistemological transformation. The formal understanding of the environment and sustainable development passed from a Marxist account of applied scientific rationality in the name of the public good, to a positivist science underpinning commodity management. Such epistemologies are not free-floating abstractions but conceptions of reality anchored into the structure of society, part of its praxis. Whilst it may be generally agreed that the political practices in the socialist period failed to live up to the social scientific propositions of a Marxist materialist account of development, the environmental modernization thesis based on technological prowess and management practices appears equally susceptible to validity challenges. Scientific accounts of social and environmental reality are clearly only partial explanations, subject to the challenges of traditional knowledge, the practices of everyday life and the distortions of political power. As a result, their predictive capacity is especially uncertain.

At the political level, improvement of the physical environment of Lodz was a crucial requirement of a symbolic break with socialism and its legacy. Just as the environment was a means to ending the socialist régime, it was also a means of legitimating the transition to capitalism. Environmental improvement through ecodevelopment was at the centre of political transformation and became a key policy commitment of the democratic governments that followed. Economic collapse and industrial restructuring facilitated this enormously, though undoubtedly the Polish state and local government devoted considerable resources to this end. However, neither before nor during transition was the environment a significant issue for the majority of citizens in Lodz. This tends to support the argument that the environment has been more important as a means to an end than an end in itself, an issue of power politics and system change rather than sustainability as some kind of objective long-term goal.

As a consequence, the environmental legitimation of democratic local government had to be consciously created through management practices. Under the socialist system, 'sustainable development' was a doctrinal affair and responsibility of the Communist Party, on which the majority of citizens could rely and perhaps at the same time continue their everyday lives on the basis of religious beliefs and traditional practices. The capitalist notion of sustainable development holds no equivalent assurances and stresses the responsibility and need for the active involvement of everyone. Rather than a theoretically inherent characteristic of development, sustainability had to be nurtured through demonstration, education, new urban imaginaries and attempts to change the conduct of people's everyday lives. However, this overt intromission through management into the field of values and personal decision-making in turn produces awareness of the logical inconsistencies, political abuses and practical dilemmas of sustainability. The battle over lifestyles seems to be a losing one when confronted with the stresses and seductions of capitalist consumerism.

Is Lodz any more 'sustainable' as a city? Whilst the condition of natural resources has improved in the short term, along with some regeneration of the built environment, the long-term trends are more troublesome. Lodz is cleaner but consumes more energy, livelier but more congested, wealthier but more unequal, and so forth. Even more so, it is the social agenda of sustainable development which appears most at risk as issues of equality, social health, participation and inclusion continue on a negative downward spiral. It is insisted in Lodz that these are temporary deficiencies which will be solved as economic growth increases, rather than universal characteristics of unrestrained capitalist accumulation.

There cannot, of course, be any absolute answer to such a question. Over the period of transition Lodz has become both less 'sustainable' after abandoning the tenets of socialism and more 'sustainable' in following the capitalist path of ecological modernization. They are two different truths of sustainability. The wholehearted adoption of capitalist urban discourse is bringing with it a rapid move towards a typically Western urban development agenda: economic restructuring, city promotion, industrial and technology sites, image rebuilding, heritage selling, the culture industry, shopping centres, new suburbs for the new rich, social differentiation, rising crime, problems of inclusion, new forms of planning, the dilemmas of greenfield or brownfield growth, infrastructure improvement, the uncontrollable rise of the car and road congestion, and so forth. The ultimate question is, therefore, whether and in what sense capitalist space is sustainable.

Chapter 7
Urban environmentalism in action 3:
defusing violence in Medellin, Colombia

Introduction

Medellin is a city of two million inhabitants and centre of a metropolitan district comprising nine other municipalities and a total population of just over three million. Located in the central cordillera of the Andes some 440 km by mountainous road from the capital Bogota, it is the centre of the Antioquia department. Colombia's exuberant tropical geography made communications difficult and fostered the development of relatively independent regional economies and strong local cultural identities. Antioquia, with Medellin at its centre, has long boasted a sense of national leadership to rival that of Bogota, based on religious discipline, early industrialization, an enterprising spirit and ethic of hard work and hard play. At an altitude of 1500 m, it is known as the city of eternal spring.

Gold mining followed by coffee production were the bases of initial capital accumulation, followed by industrialization under the relatively isolated and protected conditions of the post-war economy. In 1930, Medellin was a quiet town of 100,000 inhabitants nestling in a rich upland valley. Now it claws its way ever higher up the steep valley sides and is bursting at the seams. It underwent the typical Latin American process of rapid and uncontrolled growth from the mid-twentieth century onwards, but responded in many ways with outstanding efficiency, boasting some of the highest utilities provision rates in the continent by the end of the century, with 99 per cent of households having electricity, 94 per cent having water and sewage connections and 82 per cent having domestic telephones (Medellin Planning Department, 1999). However, in the 1980s things started to go wrong. Accumulated urban problems and industrial recession combined to produce economic stagnation, widespread impoverishment and accentuated spatial segregation (Medellin Planning Department, 1985). Latent tensions rose to the surface and then exploded with the emergence of local drugs cartels. A 'culture of violence' began to set in and the number of homicides in the metropolitan area reached epidemic proportions, with a peak of 6644 in 1991 (1 in 400 people per annum). Urban violence became the everyday obsession of all social groups (Jaramillo, 1995) and source of a perverse international notoriety as the most violent city in the world.

Two major policy areas were designed to re-establish public order and local state control. The first, most obvious and widely recognized one was overtly political. In an unprecedented move, a presidential commission was set up to address the critical social problems of the city directly, particularly those associated with the high levels of generalized crime and violence and the drugs cartels. The strategy involved a policy of political negotiation with illegal organizations (drugs cartels, urban guerrilla movements and criminal gangs) and the establishment of non-violence pacts between warring factions in the popular sectors of the city (Medellin's Presidential Programme, 1992).

The second policy area was explicitly spatial and aimed at improving living conditions in the poorer areas of the city. It focused on people's everyday lives within the context of escalating violence, disorder and desperation. Given the exceptionally high levels of public service provision and the generally decent physical condition of housing, spatial improvement was largely, though not exclusively, a qualitative challenge. In the case of Medellin we examine how, in the face of an acute urban crisis, the environment and the idea of sustainability played a key role in the symbolic reconstruction of a sense of unity and cohesion. This is not, of course, some innate quality of the environment but something that had to be socially constructed through the mobilization of meanings of the environment and their exploitation in relation to the overall urban problematic.

Urban planning discourse: environments of harmony and peaceful coexistence

In this first section we examine the general meaning of the environment as constructed through the city's development plans. The formal requirements and procedures of development planning in Colombia have undergone considerable changes over the past 20 years but a persistent characteristic has been their integral nature. City planning has always been concerned with broad economic and social development issues rather than simply physical land use, with the result that plans are as much political statements as technical documents. More recently, development policy is derived from the mayor's electoral programme and its spatial implications must take into account longer term land use or territorial organization plans. As a result, development plans provide a concise source of information concerning the relative importance attributed to the environment in urban development. More importantly for our purposes, development plans can be seen as the discursive presentation of the concepts and arguments used to construct the environment in terms of the public interest and so legitimate public action in its name.

The tragic end of modernism and its environmental precipitation

Since 1950, the city of Medellin had developed along lines set out in a master plan prepared under the guidance of Le Corbusier's disciples Paul Wiener and Jose Luis Sert. The city that emerged under the typical Latin American conditions of rapid and informal growth was not the one envisaged in the master plan but the latter proved remarkably successful in defining broad land use organization and infrastructure provision, in the formal sector at least. Formal adoption of the master plan

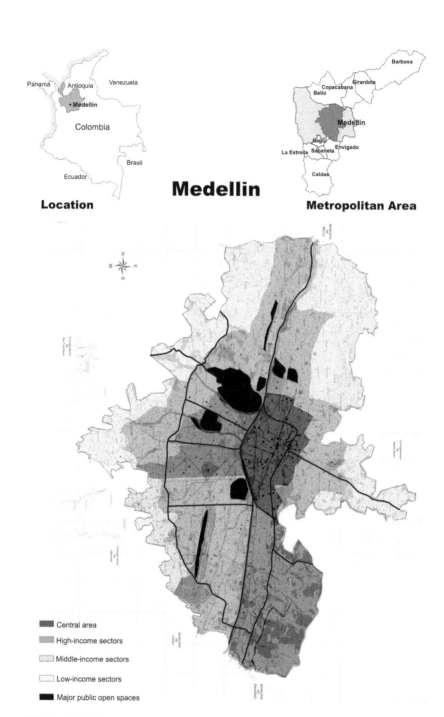

Location

Medellin

Metropolitan Area

Panamá · Antioquia · Venezuela
· Medellín
Colombia
Brasil
Ecuador

Barbosa
Copacabana · Girardota
Bello
Medellin
Itagüí
La Estrella · Sabaneta · Envigado
Caldas

Central area
High-income sectors
Middle-income sectors
Low-income sectors
Major public open spaces

7.1 Map of Medellin.

with minor modifications was effected in 1959, and in the late 1960s and early 1970s important advances were made in terms of planning regulations, transportation plans and proposals for the city centre. However, it was not until 1978 that a national framework was established for municipal planning, in the form of comprehensive development plans. Metropolitan authorities were also created a year later, Medellin being one of the first five to be set up in Colombia. Following several additional urban and regional studies, a metropolitan development plan was produced in 1985, prepared by the metropolitan planning office with the aid of international and local consultants (Medellin Planning Department, 1985).

As a formal planning document, the 1985 plan was important in that it represented the first comprehensive examination of urban metropolitan development and integrated policy formulation in 35 years, as well as the initiation of a new era in local development planning in Colombia on the basis of a newly-established national legal framework. Subtitled *Towards the Consolidation of the Metropolis*, the plan addressed what it considered the major problems: economic stagnation, poverty and the spatial inequalities of the emerging metropolitan area, and the need for new administrative structures to meet the challenges of the future.

The environment was still a distant preoccupation on the long-term horizon. As a quality of life issue, the environment received only nominal attention compared to extensive analysis on the topics of income distribution, unemployment, food prices, housing, education and health. The question of ecological imbalance was raised in five short paragraphs in relation to micro-climate modification, pollution, deforestation, water resources and environmental risk through landslides. However, the plan's scenarios ignored completely the environmental dimension of urban futures, and of the seven policy principles, the environment was announced in terms of the general need to safeguard society's long-term interests. At best, the environment appeared sporadically as a still vague and abstract concern and imprecise qualification pertaining to particular sets of proposals, especially land use, industry and transport. This timid entrance of the environment on to the city's planning stage was part of the last cry of urban modernism, not just in the sense of its pre-environmentalist conception of space and urban problems, but also in its underlying approach to planning as a centralized state activity, the positivist faith in the ability of rational analysis and expert knowledge, and insistence on universal values and goals concerning material equality and collective social welfare.

Conditions were to change rapidly. In the same year, 1985, ash emissions from the Ruiz volcano, located some 150 km to the south of Medellin, caused a huge mudslide which completely destroyed the town of Armero, with the loss of nearly 8000 lives. Medellin was not affected but the environment suddenly passed from being a long-term concern of vague strategic interest to become an urgent national issue of life and death. It had resonant consequences for urban planning. In effect, it was the most recent of a long series of natural disasters during the 1960s and 1970s (National Office for the Prevention and Attention of Disasters, 1991). The response in the municipal development plan of the following year was a limited and largely technical one. The bleak analysis of the metropolitan areas' economic and social situation was the major concern, especially given the concentration of crime, violence and social deterioration in Medellin itself. The 1986 plan attempted a more upbeat tone, identifying eight 'positive trends' in urban life: a recovered sense of credibility in public institutions and the city as a whole, renewed solidarity, new forms of employment, institutionalized public participation, fiscal recovery, lower mortality rates, more and better public transport, and improvement in the state of the environment. The rationale was not convincing and the plan was never formally approved by the city council. The most interesting aspect of the plan was its association of the environment with the general sense of violence and vulnerability affecting the city:

> A vital concern is that of citizen security, without which the panorama will remain over-shadowed by anxiety and fear. This security must be founded on state initiatives on the prevention and suppression of crime in all its forms. However, such measures will be ineffective unless accompanied by all citizens in the fight against crime: collaboration in the vigilance of public space and private property, crime reporting to the police and, above all, a change in social behaviour which has become lax and complacent on many fronts. Also of fundamental importance will be those measures designed to prevent tragedies caused by landslides and floods in the high-risk zones already identified in the city, improved disaster management procedures should such events eventually occur, a reduction in road accidents, and better fire prevention and control. To the extent that the community manages to recover a sense of tranquillity and become co-author of its own security, it will in turn be able to devote greater energy to building progress in the city and consolidate confidence in the future.
>
> (Medellin Planning Department, 1986: 39 our translation)

In the plan, citizen safety and social and environmental risk are given an associative relation, articulated to collective well-being through individual responsibility and the need for a change in attitudes and behaviour. The possible structural causes of risk (social or environmental) are marginalized, if not ignored altogether, and stress placed on individual participation in proposed solutions. The social context of such formulations was crucial. The city was entering a period of multiple crises: the local economy was undergoing a period of recession, to which the traditional manufacturing industry of Medellin was particularly susceptible; the construction of the flagship Metro mass-transit system was in grave financial difficulties, local drugs cartels were beginning to have a serious destabilizing effect on urban society, corruption scandals abounded and the credibility of traditional political leaders was deteriorating. As subsequent development plans were to acknowledge, civic self-confidence was ebbing alarmingly and a sense of vulnerability pervaded. However, at that very same moment the environment was emerging as a vector of this vulnerability, and could be usefully exploited to deflect state responsibilities and political complicity or impotence with regard to the general and critical decline in the city's fortunes.

Death, life and the environment

A year after the 1986 plan, the environmental dimension of risk was given a fatal reminder in Medellin itself. A landslide in the Villa Tina district of the city destroyed 300 homes and killed 500 people. The emotional impact on all social sectors and the city administration was enormous (Bustamante, 1987). In response to the Armero tragedy a year earlier, in 1986 the president of Colombia had established a national office for the prevention and attention of disasters, later developed as a national system for disaster management. In Medellin, the Villa Tina tragedy led to a joint programme being set up with the United Nations Development Programme in 1988, designed to identify areas of environmental risk and promote disaster mitigation and management, and the relocation of residents in high-risk zones.

The Villa Tina tragedy was the culmination of local environmental problems. Environmental risk and vulnerability had increased proportionally to the city's largely uncontrolled growth through squatter settlements and illegal developments. Between 1977 and 1988 a total of 146 landslides occurred which caused the destruction of 911 homes with the loss of 688 lives, with over 4000 people having to be evacuated (Bustamante, 1988). The scale of the Villa Tina tragedy thrust the environment to the forefront of public preoccupations and the political and technical agenda of city management. It is not surprising, therefore, that in the period immediately following the 1986 plan, the environmental question gained considerable momentum in Medellin. The publication of *Our Common Future* (WCED, 1987), had an important impact through the local media and in academic and NGO circles. Several seminars were held in Medellin, organized by universities, the chamber of commerce and the city council. In his introductory speech to one of the latter, the leader of the city council was moved to proclaim, very much in the Brundtland spirit of those times:

> Together with violence, hunger and social decomposition, the citizen of the twentieth century is witnessing the painful spectacle of the extinction of his natural environment and his own life, in a future scenario which to many seems irremediable . . . This forum will suggest priorities for action. State initiatives must be backed up through education and community awareness. Solutions are urgent and there is no time to lose.
>
> (Medellin City Council, 1989: 7 our translation)

Growing national concern over the environment was also expressed in the 1989 Urban Reform Law. Whereas the main intention of this legislation was to create legal instruments for more effective urban land management, environmental issues were central to the substantive modifications of plan content. The law obliged renewed commitment to the protection of areas of environmental or ecological importance, a reaffirmation of the need to incorporate environmental legislation in development plan preparation, and the obligation to undertake risk assessment exercises and provide land for the relocation of people living in high-risk zones. Furthermore, public space was introduced as a new key element in urban planning, defined as 'the set of public spaces and the architectural and natural elements of private property which, in their nature or use, satisfy collective needs and therefore transcend the individual interests of inhabitants'. In this way, the concept of public space was connected to the environment by including 'conservation and preservation of the landscape and natural elements of the city environment', and its defence was to be aided through the provision of community-initiated legal actions.

A further development plan produced in 1990, a hurried response to the obligations contained in the Urban Reform Law, argued in its general statement that the city was suffering 'serious environmental problems . . . upon which action must be taken in an opportune and appropriate way, in order to avoid Medellin becoming an unlivable city in the very short term'. In contrast to previous plans, it was now claimed that the environment should not be treated as a sectoral issue but considered as a vital consideration in all urban policies. In practice, the environment was treated sectorally, with two overriding considerations: first, the identification of high-risk areas and the allocation of land for relocation purposes (as required by the 1989 Urban Reform Law) and second, the problem of contamination. Studies had revealed a worsening of air quality and noise, with particulate matter and carbon monoxide levels well above permitted limits, especially in central areas. The concept of sustainable development was not included, despite the grander debate circulating in the city around this idea, but the addition of public space to the issues of disaster management and pollution control (the two previously established issues) brought the environment closer to the health and safety of all citizens. In general, the plan concluded:

> The environmental picture – one of pronounced deterioration – makes it imperative that the municipal authority . . . undertake an environmental management programme in accordance with the requirements of urban development and the quality of life of the city's inhabitants. Special attention should be directed to the following aspects: cleaning up rivers, the control of atmospheric pollution by vehicles and the future disposal of solid wastes. Similarly, emphasis should be given to basic service provision and environmental health in the marginal sectors of the city, with the aim of ensuring an adequate spatial organization and incorporating these sectors into the overall city development process.
>
> (Medellin Planning Department, 1990: paragraph 3.6 our translation)

However, whilst environmental risks were beginning to receive serious attention, social risks continued to escalate. As recorded in the 1993 plan, the homicide rates of the early 1990s were staggering, with around 5000 murders per year in the city. Violence had reached epidemic proportions, affecting all classes and groups of local society. Gilberto Echeverri, Governor of the Department of Antioquia at the time, was later to comment that when he assumed office he found the city 'frightened to death . . . most of the local elite had fled to other cities or countries' (*El Espectador*, 1997).

A crucial event in both environmental and social terms was the run-up to constitutional reform in 1991. From a political viewpoint, the most sensitive issue was extradition, solicited by the USA and fiercely opposed by the Colombian drugs

7.2 An exuberant tropical geography and mountainous setting were important factors in the construction of environmental meaning in the city.

cartels centred at the time in Medellin. The latter waged a campaign of political assassinations and intimidation, a war raged between the mafia and the police, and bombings and shootings were a daily occurrence on the streets of Medellin (Jaramillo, 1996). Such was the scale of conflict and violence that in 1990 the national government undertook the unique step of creating a special presidential programme for the metropolitan area of Medellin. The aim was to coordinate the actions of central and local government institutions in the city, obtain international finance and aid, and promote concerted actions with civil society in order to restore peace within the city (Medellin's Presidential Programme, 1992). Given the national and international scale of the problem, nine months later a further central government initiative led to the setting up of a complementary programme for the promotion of peaceful coexistence to specifically combat the drugs-related aspect of the urban crisis.

The new constitution was also a major factor in the institutionalization of the environment. The urgency of constitutional reform was precipitated by economic, social and state crises (Bejarano, 1992), but it also provided an opening for the increasingly important international environmental agenda. A constituent assembly was established, involving representatives of the main political parties, ex-guerrilla organizations and minority ethnic and religious groups. The environmental coalition did not get a member elected (managing only 0.18 per cent of the total vote) but individual academics and activists were influential in their advisory capacity to the main political groups on environmental issues (Febres Cordero, 1991). The environmental content of the new constitution was principally developed in terms of national development interests (natural resource protection, control of genetic resources, a ban on biological, chemical and nuclear arms and toxic waste disposal), ecological limits on private property and free enterprise, and state territorial organization (Rodríguez Becerra, 1992). A healthy environment was established as a collective right, and the overall effect of the constitution was to place the environment firmly in the realm of state responsibility and, therefore, into the heart of public institutions.

The Earth Summit held in Rio de Janeiro in 1992 gave further impetus to the environmental question. As part of the preparations for that conference, Colombia produced a national environmental policy document, substantially based on the inclusion, for the first time, of a section devoted to the environment in national development policy (Guhl, 1992). Criticized for being technicist and short on effective measures, it did at least address urban environmental issues in a significant way, especially with regard to land use management, clean water supply and pollution control.

By the time of the city's 1993 development plan, the combined effect of constitutional reform, innovation in national development policy and the Rio Earth Summit was to place the environment firmly on the political agenda at both national and local levels. Furthermore, the plan argued that Medellin was beginning to emerge from the worst of the violence and killings associated with its confrontation with the drugs cartels. A sense of relief permeated the city, along with an urgent awareness of the need to re-establish social harmony, rebuild social structures and restore urban self-confidence.

The 1993 plan was published with an extensive report by the city council's planning committee, which involved a heady rhetorical mixture of elements from the philosophy of law, national history and urban theory, and an extensive analysis of the social situation in the city, emphasizing problems of social inequality as expressed through spatial segregation, poverty, income disparities, high unemployment, deficiencies in education and health services and high levels of violence. In the plan itself, environmental qualities were seen as key elements in the consolidation of a sense of urban belonging and identity: 'Belonging and identity should be strengthened through the conceptual value of meeting places, open and green space, and recreation and sports areas, the concept of the neighbourhood, ecological attitudes, environmental values and the affirmation of coexistence' (Medellin Planning Department, 1993: 90). Sustainable development was introduced as one of the eight general principles, understood as development 'in harmony with the natural environment' and a requirement for the 'peace and prosperity for future generations', to be based on the 'adequate and rational use of human, natural and artificial resources'. This introduction of the idea of sustainable development into city planning involved the adaptation of the Brundtland formula to the Medellin context of violence:

> Planning in Medellin will be undertaken from a human perspective and with a sense of responsibility for future generations, through the adoption of the criterion of sustainable

development. In meeting present needs, deficiencies arising from the past will also be addressed, along with the setting of conditions which guarantee peace and prosperity for future citizens. The governance of the city will be ruled by the principles of respect for human life and all other forms of life, the integral development of the city's inhabitants, and therefore of the city itself, and will be realized in perfect harmony among human beings and between human beings and the ecosystem which surrounds us.

(Medellin Planning Department, 1993: 97 our translation)

The nature of the 1993 plan, a strategic document dealing with general principles and weak on concrete proposals, was appropriate for formally reconceptualizing the environment in relation to the city's overall problems. Encouraged by the international recognition of the environment as an integral development issue, the environment in Medellin became a device for reconstituting a sense of wholeness and harmony in a socially fractured and violent society and a means of solidifying optimism in the future. The environment became intimately associated with notions of equality, security, solidarity, peaceful coexistence, rationality and harmony: all those qualities so notoriously absent from the world of social relations in the city at that time.

These themes were to persist throughout the decade, put into action through specialist institutions and encouraged by extensive and varied programmes of environmental education in neighbourhoods, schools and the media. The dissemination of environmental values and knowledge fell on fertile ground in all sectors of the city, resonating with the personal and collective experience of social space. It mattered little that many of the more technical aspects of conventional environmental agendas such as solid and toxic waste disposal and air pollution control should splutter or fail. Enough was being talked about and done to create the sensation that the environment was a significant and meaningful area of intervention in relation to the city's particular social crisis. In the name of the environment tangible improvements were being realized in urban space and the city's social life. It was unnecessary to talk formally about a local Agenda 21; something like 'sustainability in action' was being invented locally with a minimum of reliance on international policy. Despite the fact that plans towards the end of the 1990s began to introduce a competitive edge to urban living in relation to economic globalization, the environment remained the privileged arena of reason, harmony and healthy collective living.

Restating security and new spaces of peaceful coexistence

With the change of the millennium, the new city mayor and his planning team made an abrupt abandonment of the construction of meaning in the environment. In the development plan of 2000, *Competitive Medellin*, the environment and ecology were no longer references for citizen coexistence and were replaced by public space as the structuring element of city life:

> The city is learnt and teaches in its streets. For this reason, it is necessary to feel and live the city as a scene of permanent learning and growing, not survival. We should never cease to be surprised by what we have. We should become tourists in our own city . . . Our city has beautiful laboratories for the practice of life: streets, corner shops, areas for sport and recreation, museums and other spaces for the diffusion of knowledge, parks and places for meeting and conversation. They all require a common thread to ensure their educative accent.
>
> (Medellin Planning Department, 2000: 3/26 our translation)

In an implicit abandonment of ecology for the purpose of social integration, it was argued that the greatest urgency facing the city was to find new ways of creating a better life. However, no equivalent alternative to nature, ecology and the environment was proposed. Instead, the development plan announced a moralizing 'urbanity' with some very concrete connotations:

> A new urbanity demands new behavioural values which strengthen coexistence. We cannot tolerate drunken drivers in their double condition as violators of norms and potential killers; citizens who disturb the tranquillity of their neighbours; people who are intolerant of others different from themselves; violent football fans who attack opposition supporters for simply identifying with the other team; or public officials who are specialists in making people wait.
>
> (Medellin Planning Department, 2000: 3 our translation)

This call for a 'revolution in urban culture' lost all the metaphorical richness of nature and ecology, falling prey to a routine of moral appellations and normative exhortations. The turn to public space was in part a response to changes in planning law and the extraordinary success in Bogota of the policy of 'citizen culture', a conceptually innovative programme aimed at improving urban social life through the articulation of the legal, moral and cultural modes of regulation of behaviour. Public space was the principal setting for the activation of this new agenda and enormous investments were made in urban spatial improvements which significantly upgraded the quality of parts of the built environment and radically improved Bogota's city image. However, the improvized adoption of the 'citizen culture' programme in Medellin lacked political depth and cultural foundations, whilst at the same time specialist environmental institutions in the city were deactivated in an administrative reform. In the space of two or three years the patient construction of meaning in the environment was abandoned, although the regulatory framework and residue of environmental significance is still a vital factor in specific development issues and urban politics, but now largely located outside the city administration.

On the other hand, this discursive abandonment did not occur on its own, in a political vacuum. The political climate in Colombia was becoming more bellicose as the peace process with the FARC guerrilla movement faltered and new forms of political violence and organized crime began to emerge in the city. Public frustration was radicalized by the authoritarian populism of the new hard-line president, elected on the promise of military defeat of guerrilla movements and a platform of law and order. The Medellin strategy reflected this sway of public opinion towards authority, obedience to norms and respect for order. Even so, it is not unreasonable to ask whether the environmental strategy of social regulation through symbolic meaning was abdicated or simply ran out of steam. Whatever the case, the end came in October 2002 when troops supported by tanks and helicopters stormed the Comuna 13 sector of the city to repress the public disorder and violence there, when before it would have entered with excavators, environmental engineers and tree saplings.

The institutional mobilization and control of meaning

In the previous section we examined planning policy as formalized discourse or the '"hard copy" instance of planners' discourse that planners intentionally write for public consumption' (Tett and Wolfe, 1991). Formal discourse gives the impression of a logically coherent and politically consistent set of policy propositions which came to an abrupt end in the new millennium. The situation was evidently much more complicated, and discourse analysis also involves tracing out the dispersion of often logically incompatible and contradictory discourses which constitute a discourse formation. Many different viewpoints existed on what the environment was and meant in Medellin, and exploring the heterogeneity of urban environmental discourse opens the way for the critical examination of the effects of power on the construction of knowledge and meanings of the environment, and the broad political interests inextricably caught up in talking about and acting on the environment. Also, participation in formal discourse tends to be highly restricted, so that the meanings constructed have to be mobilized through the institutions and agencies of concrete urban change in order to be introduced into the understandings of everyday urban life by ordinary citizens. With this in mind we first introduce the institutions set up in the city before developing an analysis of their function in the mobilization and control of environmental meaning.

The institutional complex

Institutional organization around the environment can be traced back to very diverse origins in the early 1970s. To celebrate the visit of Pope John Paul II in 1972, for example, the city went to great lengths to improve the image of his procession route, cleaning up and decorating streets with floral arrangements. The impact of this was such that the person in charge was declared the permanent 'green mayor' of Medellin, largely financed by local industrialists. At the other extreme, over the same decade several ONGs emerged in the environmental field with roots in popular movements and activism in the poor sectors of the city. In the early 1980s, the public utilities company initiated a long-term programme to decontaminate the river Medellin flowing through the metropolitan area, involving a huge investment in the sewage network and decontamination plants. In the local universities, groups of academics inspired by international development policy also began to take an interest in environmental issues and promote environmental initiatives in fields such as air pollution and land use planning.

It is against this background that specialist institutions were created in the late 1980s and early 1990s within the city administration. The first was the office for disaster prevention and management, after the Villa Tina landslide mentioned earlier. Then in 1992 the My River Institute (Instituto Mi Rio) was created with responsibility for the integrated management of the river Medellin basin. As both an executive and coordinating body, its governing board was conformed by the city mayor, the head planner, the chief executive of the metropolitan authority, the general and technical directors of the public utilities company and the executive director of the chamber of commerce; a community representative was later included. In this sense it acted as a discursive centre, but more important was its disseminating function. For example, its first complete management report (Medellin Municipality – My River Institute, 1997) gave a detailed account of public involvement in 1995–96 through mobile-unit educational programmes (involving 60,000 people), stream clean-up programmes (63,512 people), training and awareness (2000 people), educational workshops related to specific project implementation (33,000 people), the Friends of My River Club (1300 members), environmental education workshops (14,976 people representing 229 community organizations), a pilot scheme on didactic material (120 students), a literary competition (322 contestants), an environmental Christmas (3000 participants in ecological nativity scenes and 72,000 attendees), a recycling contest (120 groups), an annual river basin day celebration (30,000 people) as well as the involvement of groups of teachers, nuns, priests and Boy Scouts. The institute also edited publications, promotional brochures and videos, as well as mounting TV, radio and press campaigns.

These local initiatives anteceded the creation of the Ministry of the Environment and regional environmental authorities in 1993. Cities such as Medellin with a population of more than a million inhabitants could assume environmental authority within their urban areas, a function eventually given to the metropolitan body, although two different regional authorities controlled the rural part of the city's jurisdiction. By the early 1990s, an institutional 'complex' was in place with sufficient policy influence, regulatory powers and financial resources to ensure that the environment was firmly entrenched as an urban policy issue. These institutions were not only obliged to cooperate internally but also to work with the private sector, NGOs and community organizations on a daily basis and through especial programmes such as the presidential programme for Medellin and a major internationally-financed residential improvement project in the informal sectors of the city. This guaranteed the circulation of ideas and arguments and facilitated the mobilization of general meanings established through the formal planning process amongst a wide variety of organizations and across project areas. However, this was no simple process of rational decision-making, as we shall see in the following section.

The official voice of urban environmentalism

A key proposition of discourse analysis is that power operates through discourse not on the basis of some kind of superior objective truth, but rather through the institutionalized predominance of particular meanings over other possible interpretations of events. This should not be construed as a deterministic trait of discourse. As Harvey (1996) argues in relation to both his own and Foucault's work, discourse is always an 'internalized effect of other moments in the social process', an interplay of relations to material practices, social relations, power, institutions and beliefs. In this sense, discourse analysis is concerned with revealing the 'effects of truth' which dominant discourse seeks to obtain. Six 'effects of truth' characterized official discourse in Medellin. The first one – the openness and ambiguity of the term 'environment' – can be understood as a condition for the coexistence of the other five: the emergence of a species voice, the naturalness attributed to environmental problems, the assignation of common responsibility for environmental care, the establishment of legality through meta-discourse and the naturalistic conception of rationality.

Defining 'environment'
The lack of precision over the term 'environment', the set of objects and relations which it embraces, the concepts which provide the key to understanding it, have been widely commented in environmental critique. Medellin is a typical example in this sense. The environment was very rarely defined but rather its meaning was assumed or insinuated in diverse ways and directions. An early definition used in the 1986 development plan was based on the notion of 'total environment' understood as the 'sum of the natural environment, human populations and cultural aspects' and applied analytically in terms of flows of energy and waste between urban centres and their surrounding regions. The totalizing scope of 'environment' remains a constant feature, as in the following example:

> The term environment is understood to mean all that which surrounds man: the natural, technical, political and social elements, and the relations between them. That is to say, the environment is the systemic reality which arises from the articulation of socially-organized mankind with nature, through the process of development.
>
> (Medellin Municipality/UNDP, 1996: 10 our translation)

Such all-encompassing conceptions of the environment provided the means for highlighting more concrete and specific dimensions or objects according to particular events and conjunctures. In the 1980s, the environment was associated with natural resource systems, pollution and imbalance in the form of disasters. The 1990s saw an increasing influence of global environmental discourse, both directly and through national policy and legislation. New arguments emerged concerning a healthy environment as a constitutional right, the value of ecological systems, their management requirements, educational demands, and so forth, so that by the mid-1990s sustainable development ('solving past errors and generating conditions which guarantee a peaceful and prosperous life for future generations of the people of Medellin') was used to fully integrate the environment into urban management processes. The relative autonomy of the environment was never made clear beyond the explicit or implicit reference to the natural components of the biosphere: air, water, soil and ecosystems.

The neutral voice of the species

A common feature of environmental discourse is its reference to a critical point in history which threatens the balance of life on the planet earth. As a consequence, one particular contribution of environmentalism is its reaffirmation of human beings as natural entities, if no longer immersed in the natural, ecological organization of life, at least ultimately dependent on natural biophysical cycles and vulnerable to the effects of their radical modification. On this basis a vigorous voice emerged through environmental discourse in Medellin: that of humanity as a biological species. The 1985 plan, for example, states among its objectives the satisfaction of 'vital necessities and human biological requirements' and 'ecological balance'. This, furthermore, was a question to be addressed in law:

> The rights and duties of man can no longer be considered independently from the obligation of present generations to conserve our natural heritage, the basis of development for future generations. The conception of law has to detach itself from its short-sighted egoistic productivism to be able to understand its responsibilities to future generations. Life is not a momentary pleasure, but rather a process that has been constructed and perfected over millions of years, and which man cannot disturb with impunity. The destiny of man is tied to life systems.
>
> (Gaceta Constitucional, 1991; No. 46: 3 our translation)

The establishment of the biogeneric voice through the discursive strategy of evolutionism is far from neutral in its effects. In the above statement, for example, this reinsertion of human life within an evolutionary perspective carries with it both a moral threat and a challenge to the autonomy of human society. It bears a new kind of puritanical anti-hedonism ('life is not a momentary pleasure') and an assignation of responsibility; a responsibility which, moreover, is a generalized one pertaining to human beings as a species category. This evolutionary perspective avoids the question of causes and social structures which would compromise its naturalistic and scientific authority. With respect to disasters, for example, the species voice is employed in relation to a metaphysical rupture with the cosmic wholeness of existence:

> *Mankind*, part of Nature: The harmonious existence of mankind with itself and the planet depends on the awareness and perception we have of the wholeness of the universe. The philosophic basis for the prevention of disasters should be the mankind–nature relation, as being that element which provides meaning to individual being, community being and that great being which is the Earth itself, and which expresses itself in life. From this awareness will spring processes and attitudes in accordance with our planet.
>
> (Medellin Municipality/UNDP, 1996: 9 our translation)

This direct, explicit use of the species voice frequently supplants the more usual use of the neutralizing device of the passive grammatical voice. Causes and explanations are avoided since everyone is biologically implicated and public institutions can discursively represent the common interest through the articulation of a universal biological condition. The overall discursive effect of the species voice is to provide a new form of expression of human unity which precedes social structures, sidesteps causality and diverts attention from the widening inequalities and injustices of the social conditions of existence.

Common responsibility

Closely associated with the discursive mode of the species voice is the notion of common responsibility in redressing environmental ills. This does not necessarily imply equal and similar responsibilities in terms of requisite actions, sacrifices or changes of behaviour. Indeed, the species voice is discursively employed to consolidate a sense of collective responsibility and, within it, assign particular individual or group duties. Common responsibility may be evoked through the idea of something lost or shared pleasures of previous times which have vanished without knowing exactly when or why, and frequently recalled with nostalgia. Typical of early environmental concern, this sense of collective forfeiture is illustrated in this excerpt from the introduction to an otherwise highly technical study of the river Medellin:

> In bygone days, as we will see throughout this monograph, the river washed its bed with clean water, the old inhabitants of the town bathed in it and fished the famous 'sabaleta', once a staple part of the Medellin diet. There was still sufficient water for the navigation of light craft, which used to carry building materials and food from the south for consumption in the city.
>
> (Medellin Municipality – Empresas Publicas, 1981: 9 our translation)

This sense of common loss is translated into a common responsibility with respect to remedial action. Romantic nostalgia gives way to duties and obligations, as discursively formalized through the Brundtland Report and reinterpreted in the Colombian context, for example, as follows:

> To suppose an exclusively political treatment of the environmental crisis would demonstrate an ignorance of the fundamental role which society as a whole has to play in the preservation of our common heritage. For this reason, the proposals and formulae put forward by our party contemplate both aspects of this responsibility: on the one hand, *government commitment* to the design and implementation of a coherent and appropriate environmental policy and, on the other hand, the *duty of all inhabitants* of the country to respect, defend and protect the environment.
>
> (Gaceta Constitucional, 1991, No. 94: 7 – original italics our translation)

For these abstract duties to be fulfilled through concrete actions, citizen awareness must be consolidated and mechanisms put into operation. Environmental education and knowledgeable participation become the order of the day. Just as the state of the environment needs permanent monitoring, so does the state of knowledge of the participant population and its ethical commitment to ecologically benign behaviour. However, the notion of common responsibility converts ethical concerns into behavioural norms to be cultivated, evaluated, regulated and enforced.

Naturalness

Notwithstanding the development polemic concerning the relation of poverty to environmental problems, disasters in Medellin were continually constructed as 'natural' phenomena inherent to human existence. If the mythical power of the gods could no longer be appealed to, then allusion was made to the deep forces of geology, the surface energy of geomorphological processes or the unpredictability of technology. Risk and disaster were held to be integral components of the human experience, unconnected to historical processes and social structures. The creation of this effect was achieved in various ways. One route was via the deep ecology trend in environmental thought and the cosmological vision of mankind within the universe. A similar discursive strategy was the poetic presentation of humankind's relation with nature. The final document of the disaster management programme, a glossy promotional publication designed for international consumption, contains ten quotations, of which seven are from poets, writers and philosophers. The following, from the Colombian poet Leon de Greif, accompanies a full-page photograph of a four- or five-year-old boy carrying a heavy brick through a post-disaster landscape of rocks and rubble, inset into an abstract background image of primeval colours:

> Nobody heard him, only the stones
> monuments of ancient ages
> (Medellin Municipality/UNDP, 1996 our translation)

The discursive resources activated through the cosmic and poetic imagery of nature and the environment contrast strongly with the highly technical, management-based approach to disaster prevention contained within the report. It leaves the question of who speaks drifting in the air; certainly there is no voice of the affected and the destitute. In his presentation of the same document, the city mayor introduces the idea of environmental risk as something to which we all are 'exposed', like sunlight or gravity, in which case recourse should be made to the protective hand of government and the 'appropriate technical measures'.

Furthermore, if risk is a natural, inherent part of everyday life within any society or culture, then it can also have a mundane dimension without losing its essential naturalness. In the 1996 plan, for example, environmental insecurity is held to be due to natural risks, undesired effects and population overcrowding in inappropriate areas (Medellin Planning Department, 1996: 99). Disasters are attributed to irresponsible behaviour, bad management or simply bad luck. Discursively, however, risk in Medellin was associated with fate more than the reflexivity of technology-led risk society. Here, the interplay of environmental discourse with other belief systems is undoubtedly important. Medellin is a strongly catholic city where religious practice forms an important part of everyday life. Urban culture is no stranger to techno-scientific bureaucratic systems or a culture of progress, but religious tradition meant that fate could provide a deeper and emotionally more persuasive explanation of events.

Legality/status
The authority of the environmental discourse of specialist local institutions constantly sought legitimation through reference to global environmental policy. This draws attention to what Foucault calls 'enunciative modalities' or the questions of who speaks, from which institutional sites and on the basis of what claims to competence and knowledge. As much as prestige and administrative authority, at stake is the status of the environment as a problem and the legitimacy of the objects and concepts of environmental discourse. This effect operates through several different channels. The most obvious one concerns the major international declarations and conferences (the Brundtland Report, the Rio Summit, the Istanbul Habitat II conference, and so forth) which institutionalize an environmental meta-discourse. Participation in such events gives a status which is discursively exploited through, for example, Colombia's document presented to the Habitat II conference in Istanbul:

> Colombia's participation in recent world meetings on human settlements, such as the Cities Summit in Curitiba in 1992, the [Sustainable] Cities Conference in Manchester in 1994, and especially the World Summit in Rio de Janeiro in 1992, generated commitments in our country such as the inclusion of urban and environmental dimensions in the national development plan.
> (Ministry of Economic Development, 1996 our translation)

A second level concerns the way in which single issue international programmes are used to guarantee the status of local urban environmental problems. The question of disasters is thus discursively contextualized in the quasi-legal forum of the United Nations, as well as a contractual agreement with that organization:

> In accordance with this policy [of disaster prevention] and coinciding with the United Nations declaration of the 1990s as the Decade for the Reduction of Natural Disasters, the mayor is working with the help of the UNDP and various municipal organizations to study and analyse the landslide, flood and earthquake risks to which the city is exposed.
> (Medellin Muncipality/UNDP, 1996: Presentation our translation)

Alternatively, the international networking of cities may provide the discursive resources for maintaining the status of the environment as an object of local urban intervention:

> A pressing obligation was acquired by the municipality, and particularly the My River Institute, through the declaration of the mayors' conference on the environment held in

Marseilles (France) last March: 'The greatest importance needs to be given to information and education at all levels, particularly the young, so as to encourage disaster prevention and awareness'.

(Medellin – My River Institute, 1995: Presentation)

The discursive status acquired through international agreements and transcribed in local policy documents is then introduced into both national and local urban regulations. In Medellin, new planning regulations in 1997 had 66 of the 250 articles referring to the environment, as compared to 23 of the 560 articles in the earlier regulations of 1990. Discursive status and the legal mode of enunciation, with all its institutional requisites and apparatus, are mutually reinforcing.

Rationality

Notwithstanding the widespread environmental critique of instrumental rationality, urban planning and management are founded on the rational (open, explicit and logically argued) use of public resources in function of politically defined aims. However, in practice rationality in public administration in Medellin was undermined by the imposition of the logic of the market, the difficulties posed by social fragmentation and manifestations of corruption, incompetence and inefficiency. The violent turbulence of social life proportioned an additional sensation of chaos. It is in this context that the environment provided a plausible alternative account of rationality as a legitimating facet of local politics and urban interventions. Within environmental discourse, to act rationally is to condition social practice to the laws and regularities of natural resource systems. At the same time, the concept of a sustainable environment discursively introduced the future and with it re-elaborated the notion of politics as a long-term social project which addresses a sense of tomorrow:

> The second clause sets out the criteria by which the state should undertake its [environmental] responsibility. In effect, it establishes that it is the duty of the state to promote the use of natural resources in terms of development and an improved quality of life for present generations, and that, at the same time, their use and management should be *rational*, in such a way as to maintain the potential of the environment to satisfy the needs and aspirations of future generations.
>
> (Gaceta Constitucional, 1991, No. 46: 4 – original italics our translation)

In a similar vein, the 1993 development plan states as one of its objectives: 'To ensure, through the policy of sustainable development, that an adequate and *rational* use is made of human, natural, and artificial resources, and of all other resources belonging to the city' (Medellin City Council, 1993: 97 – our italics). The 'rational' is rarely explained but rather, as indicated above and in previous sections, achieves its effect through constant repetition. By this means, the urban environment becomes reinforced as the discursive domain for the re-establishment of the rational action of society upon itself, now through the incorporation of the laws of nature, and in this discursive shift the environment becomes a vital legitimating device for public institutions and a central ground of local politics.

Heterogeneity and the containment of conflict

The official voice of urban environmentalism in Medellin did not go unchallenged. As confirmed in a series of interviews, environmental organizations questioned the narrow focus of environmental policy and its limited scope for bringing about significant social change towards sustainable development. Labour organizations and NGO activists working in the popular sectors of the city rejected the narrow definition of the environment as being about natural resources, and took the environment into the home, the neighbourhood and the workplace. Left-wing politicians contested the cosmological and ecological arguments and insisted on explanations around the destructive nature of neoliberal development and unfettered capitalism. Community organizations questioned the technical competence and politicized management of specialist environmental institutions. The importance of the environment with regard to the quality of life was discredited in relation to increasing urban poverty, enormously divergent knowledge bases were brought into play when talking about the environment and public institutions themselves were prone to disagreement over policy approaches and priorities.

In other words, the 'effects of truth' of official discourse required continual reassertion. Discourse analysis regards this as constitutive of all social discourse where claims to truth and the exercise of power are being fought out. Discourse analysis rejects the existence of discursive unity through a permanent and unique object, and conceives a discursive formation as the 'space in which various objects emerge and are continually transformed' (Foucault, 1989). This division and dispersion of objects and the power interests thereby implicated is, precisely, the interest of discourse analysis.

The control of dissent

How could it be, it might be insisted, that this apparently powerful and wide-ranging discontent with, and opposition to, local environmental action in Medellin should amount to so little in urban policy terms? The environmental performance of the local authority was criticized from technical, social, economic and political angles, yet public institutions and their policies became more robust. Philosophy, capital and class, the state, realpolitiks, equality and justice arose as major foundations of dissent, yet urban environmental policy remained firmly entrenched in the same set of basic arguments, material objects and spatial phenomena. The analysis of the interplay of diverse discourses provides some answers in three important senses.

In the first place, by discursively delimiting controversy to a reduced range of objects and themes, conflict over the environment was restricted not just to certain issues but the way in which those issues could be legitimately talked about. The scientific-technical description of environmental phenomena imposed itself as a discursive condition for effective participation in urban environmental debate, isolating and subjugating those people expressing environmental concern through non-expert knowledge. Second, the truth effects of dominant discourse absorbed opposition into its own conceptual structure. Thus, confrontations over the legitimacy of the objects and actions on the environment became unwittingly transposed onto the frameworks of legality; issues of social justice were submerged by the all-embracing notion of common responsibility; political dissent echoed hollowly in the chambers of the species voice; social reform withered in the face of ecological rationality; and structural critique faded against the forces of the ahistorical naturalness and inevitability of the environmental problematic. Third, 'truth effects' mutually reinforce one another. Thus, for example, the ecological sense of rationality employed in official environmental discourse marginalized alternative (romantic, anthropological or revolutionary) rationalities, at the same time as it resonated with the species voice and the idea of the naturalness of environmental risks. Similarly, the species voice of dominant discourse supported the idea of common (biological) responsibility, at the same time as it was complemented by that of legality, as embodied in the notions of equality before the law and the universality of the meta-discourse of the global environmental statements and policies.

Consensus around the official voice of urban environmentalism was established not through rational argument or significant political change but through the acceptance of general interpretations. The effectiveness of hegemonic environmental discourse, its capacity for change and closure, can ultimately be illustrated in what might be called the demise of the radicals. As a member of a popular sector NGO recounted, the reformed criminal organizations and reinserted guerrillas found themselves ending up doing exactly the same things as public institutions and local business: tidying up rivers, planting trees and recycling waste, converted by environmentalism into the unlikely new urban housekeepers of a torn urban social landscape.

The containment of social conflict

The interesting point in the case of Medellin is the way in which environmental discourse contributed to the management of social conflict. Both the universal nature of the experience of natural systems and their qualitative connotations for economic and individual well-being were understood as providing a socially significant spatial articulation of urban differentiation. More specifically, the environment was widely held to constitute a medium or referent through which a new social contract might be forged to replace the violent confrontation of social groups which had dominated social life. This confluence of beliefs had its origins in the most diverse theoretical perspectives and argumentative strategies. Thus, for example, a representative of a business-based NGO employed economic theory and the rationality of the market:

> I ask myself whether, if all these young people who commit crimes due to unemployment were to manage a system of parks, trees and woods, we would solve not only the problem of urban green space but also the problems of violence and insecurity arising

from a lack of other opportunities. The city is modernizing its industry and that means fewer jobs; in other words, these young people are not going to have work. So perhaps thinking creatively, we might be able to resolve the environmental problem whilst at the same time contributing to the solution of economic and social problems.

A spokesperson for an environmental NGO working in the popular sectors of the city situated the question at both a symbolic and practical level:

It might sound utopic but I believe that new social pacts can be built around the symbolic value of water, for example. Which social sector doesn't have an interest in water as the element which permits life in society? From pacts around the use of the natural base of a complex city such as Medellin, to the integral responsibility of companies, their self-responsibility, it should be possible, for example, to pact real transformations in systems of production that are both economically more efficient and ecologically sustainable. I see the environment as a place for pacts, concerted action and across-society consensus, for who is unaffected by the inadequate use of resources? But that's also at the bottom of environmental conflict. I think that social conflicts have a lot to do with conflicts over the use of land and its resources.

A similar proposition relating to the social imagination was put forward by a representative of a labour organization:

I do believe that [the environment can help solve social problems]. I think that's clear concerning employment, but with regard to wider social problems, I don't see a cause–effect relation, because our social problems are very complex and any unilateral strategy, however profound, tends towards failure. Various strategies must be devised that address the problem from different angles. I think it [the environment] can help change certain perceptions and modify the 'imaginaries' of people.

In a complementary fashion, the environment was also seen as a practical referent for overcoming spatial segregation and the formation of ghettoes, as another popular sector NGO member stated:

There are ecological, pacifist groups which see their mission as one of breaking down the barriers that urban violence has erected by means of walking the city . . . they are the groups which are most aware of the relation between peace and the environment. This is a city which is impossible to walk, because this 'barrio' belongs to such and such group, this neighbourhood is under the control of another one, but nature belongs to nobody, so they say 'we're going to walk it'.

From a cultural perspective, the environment was seen to be a source for the restoration of a system of values destroyed in the descent into violence, killings, corruption and social disintegration. A representative of the business-based NGO commented:

For years the culture of Antioquia and especially Medellin has been losing values . . . prob-ably the violence in many sectors deriving from the drugs cartels liquidated remaining traditional values, such as respect for the rights and life of others. I believe that the ecolog-ical movement, which is a worldwide one and is reflected in Medellin and its inhabitants, is a good opportunity to recapture lost values, not throughout the city but in many sectors, concerning respect for life, not only human life but also for other living things.

From a radically different political perspective, a similar effect was formulated by a left-wing councillor, but from an inverse argumentative strategy:

I give priority to culture [over the environment]. Unless we improve the cultural dimen-sion, it will be difficult to solve other problems. A cultured person is someone who doesn't attack nature or the life of his fellow man.

Finally, this argumentative strategy was expounded by the leader of the city's strategic planning team:

I believe the city needs different referents to solve its problems of peaceful coexistence, and I'm absolutely convinced that if the city advances with its strategy of social integra-tion through an environmental perspective . . . that there's a real opportunity, acting from an environmental basis, to solve the problems of conflict and violence . . . Without doubt a better citizen can be constructed through culture but it also has to be possible through urban actions, particularly environmental ones.

The analysis of discourse developed in this section has tried to fix attention as firmly as possible on that somewhat elusive Foucauldian level of analysis: the discourse formation. The temptation to slide into other complementary levels of analysis – explanation, origins and causality, functional relations, conventional poli-tics – is ever-present in the effort directed primarily at the elucidation of the anonymous unwritten rules of what may or may not be said about the environment in particular circumstances. The intention has been to illustrate, albeit through a limited range of discursive 'events', how the problem of violent social relations regu-lated environmental discourse in Medellin. The amplitude of urban violence and insecurity were such that the conditions of environmental discourse were set out not in terms of a 'discourse coalition' for the imposition of a particular under-standing in function of private interests. The conditions of emergence of urban environmental discourse in Medellin were defined by a crisis of social relations in general, upon which hung the legitimacy of local government and the established legal and moral order. In the following section we will explore some of the non-discursive conditions of urban environmentalism and how an abrupt transformation occurred around the turn of the millennium. In other words, we will examine what Foucault describes as the interplay between discourse and the primary or real, and secondary or reflexive levels of analysis, or to what Harvey formulates as other 'moments' of social relations.

Socio-economic transformations and non-discursive practices

Once the environment had been defined as an object of expert knowledge and established as an issue of vital popular concern, the conditions of emergence of that object tended to retreat into the background as the spotlight of attention grew in intensity, fading into the opaque background behind the glare targeted on the principal player. This section turns up the main lights to examine the environment in a theatrically less dramatic but socially more encompassing light. The metaphor of the theatre is appropriate in the sense of a managed recreation of spatial and temporal relations. The environment as main character does not deny relations with other themes and objects such as economic development, technology, social well-being – how could it? – but schematizes and abstracts them on to a backdrop; they rarely appear in their full starkness and immediacy. It is time to go backstage and examine the conditions of production of *The Environment* in Medellin.

In the previous section we indicated how environmental discourse tended to dehistoricize itself, through cosmological and evolutionary references to time and nature, arguments concerning the naturalness of environmental problems or by simply focusing on the technical agenda of natural resource management. Discourse analysis is a useful way of recovering the historical conditions of emergence and the play of power. Foucault was concerned with time in the sense of the deconstruction of coherent linear narratives, so that his treatment of the question of temporality is an insistence on discontinuity (in the formation of knowledge and social practices over time), irregularities (inherent in that construction of knowledge over an uneven temporal surface), and disorder (behind the 'visible façade' of knowledge systems). He argues (Foucault, 1989: 83) that a discursive formation is 'no stranger to time' in that it 'does not play the role of a figure that arrests time and freezes it for decades or centuries; it determines a regularity proper to temporal processes; it presents the principle of articulation between a series of discursive events and other series of events, transformations, mutations, and processes. It is not an atemporal form, but a schema of correspondence between several temporal sites'.

We have already suggested that the practical application of the 'principle of articulation' of discursive and non-discursive events can be given more concrete form by Harvey's dialectical approach to understanding social change. Even so, the analytic scope remains enormously wide and open-ended. It is here that the concept of neoliberalism can be usefully applied to narrow the focus and centre attention on the function of the environment in the regulation of economic and social transformations at the city level.

Economic restructuring and social deregulation

Environmental discourse emerged during a period of the dismantling of national economic protection and the restructuring of an economy already facing severe difficulties. The traditional manufacturing base of Medellin was vulnerable to the first phase of exposure to international competition in the 1980s, when over 11,000 jobs were lost in the formal manufacturing sector (Medellin Planning Department,

1996), and the opening up of the national economy and intensification of international competition in the 1990s saw some uneven growth. Large local firms protected themselves from take-over by forming a 'syndicate' through which each company held shares in the others, but overall employment in the manufacturing sector continued to decline. Although subject to cyclical variations, in the 1990s and into the new millennium the city's economy saw a shift to the financial and service sectors, the declining relative importance of the formal economy and labour market flexibilization in terms of reduced job protection, an attack on collective wage bargaining, the generalization of temporary employment contracts and reduced compensation for unsocial work hours. This general informalization of a low-wage economy, with complex links to the illegal sector, simultaneously expressed both a cultural tradition of ingeniousness and tenacity and a socially fragmenting and permanent struggle for work and income (Betancur, 2004).

The social consequences were typical of neoliberal development in terms of growing poverty and inequality. In Medellin, as in the country as a whole during the 1990s, job creation lagged behind population growth, unemployment rose from 12 per cent to 20 per cent, and what net employment growth did occur was concentrated almost exclusively in the informal sector and lower end of the job market, with incomes mostly below the minimum legal wage. For most people, work became increasingly low paid, unstable and unprotected. The percentage of households with unsatisfied basic needs increased over the final decade of the century; although the Gini index of income distribution improved marginally but remained amongst the highest in Latin America at 0.52 in 2000 (Brand and Prada, 2003). The latter anomaly can probably be explained by the illegal economy based on the drug trafficking which introduced vast amounts of money into the local economy. On the other hand, large numbers of displaced persons moving into the city from the war-ravaged countryside, especially towards the end of the decade, pushed urban poverty beyond normal limits.

Two parallel but not disconnected economies arose defining the opposite poles of local economic identity: a modernizing formal economy based on manufacturing and services, and a highly lucrative illegal economy based on drug trafficking. Both were export oriented, redefined the parameters of people's material practices in terms of income generation, consumer habits and lifestyles, and precipitated the end of the traditional, rigid and paternalistic organization of economic life. Medium and large locally-owned companies ceased to be the distinctive element of economic life in the city as they transformed into more remote and sophisticated organizations. The routes to individual survival and wealth acquisition became diversified and unstable, and people's consumer and leisure experiences became amplified and fragmented.

The decoupling of local business from the urban life produced a vacuum for the city administration. Traditional city fathers retreated from political and civic life, and the regulatory function of an organized economy and stable employment became weakened in an increasingly informal and illegal economy with astronomically high levels of violence. Unable to solve the structural causes behind such violence, the environment emerged as a sphere in which its manifestations could be politically neutralized and administered.

Power/discourse and the strategy of competitiveness

Despite the declining importance of the organized economy in the reproduction of urban social life, the business sector achieved an important recuperation of control over urban policy in the latter part of the 1990s through the competing discourse of competitiveness. This was promoted first and foremost by the national government. Studies undertaken by Michael Porter's Monitor Company on the competitiveness of the national economy and its principal cities, including Medellin, led to the setting up of regional advisory councils on international trade and the formulation of strategic regional export plans. The idea was to focus attention of urban and regional economies on export markets and encourage the public sector to include business and competitiveness in local development planning.

The implications for the design of strategic urban development policy in Medellin were considerable. Local business organizations, led by the chamber of trade, assumed the technical coordination of the regional export strategy which was in turn a major input into city and regional plans. The regional council on competitiveness, again business-led and set up in 1996, coordinated the formulation of a metropolitan strategic development plan and was hugely influential in the preparation of the subsequent city plan called *Competitive Medellin*. The effect was to ensure that the city's vision became business-led, outward-looking and market-oriented. Major local projects were designed to support business performance through better infrastructure (access to the airport, conference centre, cultural centres to improve the city's image and place marketing opportunities), economic development programmes (clusters, new sectors such as call centres, medical services, design and fashion) and public policy on education, training and workforce preparation. Although strategy and plan preparation formally involved a wide spectrum of the city's institutions and social organizations, local business was setting the agenda. This new style of economic governance shifted the centre of urban policy making away from the democratic nexus of the city council and onto non-representative committees and work groups, and even the mayor's office lost the leading edge in decision-making (Brand and Prada, 2003).

The point to be emphasized here is the gradual discursive displacement of the environment by competitiveness towards the end of the millennium. The business sector possessed the market knowledge upon which its own competitiveness depended and used public–private partnerships and participatory planning exercises to circulate this self-interested business discourse amongst other city actors and create understandings of it in terms of the collective interest. It is not just that the environment became a secondary issue in urban policy; competitiveness established a whole new set of meanings for urban life. The environmental themes of interdependence ceded to the principle of competition, life-world values were replaced by material ones, and so forth. Furthermore, the objects and themes of urban competitiveness were infrastructural, designed primarily for external purposes and elitist consumption. Cooperation between the different sectors of society was a formal proposition of competitiveness but this integrative function was denied in practice by the exclusionary nature of the economy. Private sector power was re-established through discourse and against the overwhelming empirical evidence of the limitations of a competitive business sector to significantly improve the lives of the urban majority.

The reform of welfare and the rise of repression

A wide-ranging institutional reform also took place. On the one hand, local government reform initiated in the mid-1980s saw a significant shift of responsibilities from central government to municipal authorities, ranging from the democratic election of mayors (previously appointed indirectly by the country's president) to the decentralization of education and health services. On the other hand, this surge in municipal power was countered by strict central government controls over local government spending and management practices, culminating in the new millennium with severe limitations being imposed on public sector bureaucracies and average cutbacks of around 30 per cent in local government employment. The privatization of publicly-owned companies was an uneven and incremental affair rather than sweeping change, and in Medellin the privatization of the large, profitable and symbolically important public utilities company (which owns and operates water, sewage, electricity generation and distribution, gas, telephone, cable TV and Internet services) was fiercely resisted.

Nevertheless, considerable changes did take place in the institutions and delivery of welfare at the national level with important implications for cities (Garay, 2002; Restrepo, 2003). The principle of competition was introduced into public services and their management, with subsidies being gradually dismantled in the areas of basic services and public transport. Tax reforms favoured private capital and the wealthy whilst public revenue turned towards the socially regressive valued added tax. Reforms to the health service failed to effectively reach the urban poor, so that by the beginning of the new century it had become a necessary and common practice to instigate legal action in order to get hospital treatment. Both decentralized national and local authority institutions for house building for the lower and middle classes were scrapped, replaced by financial systems of demand-side subsidies for the poor, insufficient to meet the needs of increasingly insolvent households, and a scandalously flawed mortgage system for the better off (Brand, 2001a). In other words, modern welfare – always a precarious affair for the urban poor but once minimally accessible through systems of political patronage and state provision – became increasingly unstable, privatized and individualized; public spending focused inadequately on the very poor and middle-income groups were confronted with intensified economic pressures.

Against this background, there were just two areas of institutional expansion of the state. The first one, as we have already seen, was in the field of the environment, especially with the creation of the Ministry of the Environment and regional authorities in 1993, along with a host of local initiatives such as the My River Institute. Public intervention on the environment became increasingly important in providing a sense of spatial welfare, spending on the environment was funded through new property-based taxes and pollution charges, so that by the end of the century environmental institutions had become powerful local actors (Brand and Prada, 2003). In the case of Medellin, the metropolitan body as environmental authority became increasingly important in terms of discretionary budget spending, with additional powers over general and detailed development proposals. Spending on the environment, though still modest in absolute terms, was increasingly significant given the spending cutbacks in traditional welfare, especially in large cities such as Medellin.

The second area of institutional expansion was law and order, especially towards the end of the 1990s. This was given impulse in 1998 with a largely military US aid package (Plan Colombia) focused on the drugs trade and escalated following the collapse of the peace talks with the largest guerrilla organization, the Revolutionary Armed Forces of Colombia (FARC) in 2001. Presidential elections the following year resulted in a 'war on terrorism' and further increases in military spending. In 2002, the Minister of Justice and the Interior famously bracketed ecologists with terrorists, the Ministry of the Environment was fused with housing, Colombia's previously strong presence in international events dwindled at the Johannesburg conference, the regional environmental authorities were accused of restricting private enterprise and the massive chemical destruction of illicit crops through aerial dispersion was extended to national parks. If, in the 1990s, the decline in orthodox welfare provision had been compensated by its reconstruction in the environment (Brand, 2001b), in the new millennium attempts at peaceful coexistence, solidarity and the rational administration of social problems through the environment began to be overtaken by a new régime of law and order, authoritarianism and militarization (UNDP, 2003; Plataforma Colombiana de Derechos Humanos, Democracia y Desarrollo, 2003). This political and institutional shift reshaped urban discourse and the place of the environment within it.

Spatial representation

The discursive construction of meaning in the environment was also mobilized through spatial representation which gave concrete form to the metaphors of sustainable development in the city (Brand, 2000). The themes of environmental discourse such as rationality, harmony and peaceful coexistence in urban social life required expression in the spatial order of people's everyday lives. Given that violence, both social and environmental (disasters) was concentrated in the popular sectors of the city, spatial interventions were directed principally but not exclusively at those areas. It was not simply a case of addressing functional problems related to health and risk; symbols were also needed which could extend environmental meaning to the population as a whole. Symbolic meaning can not be manufactured in isolation but has to resonate with pre-established cultural understandings. In the case of Medellin there was already a cultural tradition of 'good housekeeping', with the city priding itself on being clean and well-kept with efficient public services, regular refuse collection, good street cleaning and the impeccable maintenance of house fronts. This was a cultural tradition to be exploited by modern environmentalism.

In the popular sectors of the city the My River Institute undertook environmental works centred on the river system by cleaning up streams, undertaking civil engineering works, tree planting and educating local residents on the importance of respect for natural systems and disaster prevention. More importantly, these environmental works were significant spatial interventions in sectors of the city that for years had been ignored by the city authorities except for basic service provision. Housing densities had increased enormously but spaces for education and health services lagged way behind, and provision for open space and recreation was almost non-existent. Stream management and disaster prevention became a means for creating public open space and improving the spatial organization of the crammed conditions of many popular sectors. This was something like 'sustainable development in action' with a direct impact on the qualitative experience of the urban poor.

At the city level, urban greening was extended and generalized. Initiated in the 1970s as a decorative act for the visit of Pope John Paul II, tree planting was

7.3 Improvements to the densely populated poor sectors of the city have been achieved through spatial interventions based around river and steams.

7.4 The river Medellin is the central axis of the 'multi-modal transport corridor', here decorated for the city's Christmas celebrations.

systematically promoted in the 1980s and 1990s under the impulse of ecology: urban vegetation would stabilize steep slopes, improve micro-climates, decontaminate the air, protect biodiversity and provide sustenance. An estimated 650,000 trees were planted in the period 1985–98 and millions more were planned for the following years (Medellin Planning Department, 1999). Tree planting took place in public spaces (main roads, streets, parks and any piece of open space) which meant a big visual impact. Whilst the dove had become the symbol of peace in negotiations with guerrilla movements and organized gangs operating in the city; the tree became its equivalent in terms of urban space.

The symbolic importance of the environment is most dramatically illustrated, however, in the case of the river Medellin traversing the city. In the 1950s the river had been straightened, culverted and gradually converted into the straight-jacketed axis of a 'multimodal transport corridor' containing the main spine of the road system, railway and metro, high-tension electricity cables, and gas and oil distribution pipelines. This monument to modernist functionalism was extremely efficient but equally grey, lifeless and polluted. This affront to postmodern environmental sensibility was to undergo a truly remarkable transformation. The My River Institute began by tidying up the narrow green spaces by the river and planting flowers along a new pedestrian route along a short stretch near the city centre. Quickly denounced by environmentalists as mere cosmetics, little by little the river corridor was colonized and re-signified. In a symbolic gesture of great importance the My River Institute built its cultural centre on a bridge over the river and began organizing boat races, walks and festivities. Sections of the motorway were closed off for use by cyclists, skaters and joggers in evenings and Sunday mornings with massive use, the river environs became the major centre for Christmas celebrations, permanent balconies were built overlooking the contaminated waters, and so forth. Defying all modernist functional logic but exploiting environmental meaning, what is still an urban highway has become one of the most important public spaces in the city.

Conclusions

In the case of Medellin we have argued that the environment was discursively constructed in response to the crisis of violence which began to afflict the city from the mid-1980s onwards. In the face of disintegrating social relations and a destructive human order, the environment became a metaphor for reconstructing a sense of the value of life, common interest, solidarity, rationality and an underlying legality based on nature's values, biological existence and the laws of ecology. Scientific and technical knowledge played only a supporting role, as did conventional sustainable development policy. Rather, cultural tradition and environmental disasters within the city made the metaphor a plausible one for addressing the risk, precariousness and uncertainty of the general urban conditions in Medellin: sufficiently real as to address concrete urban problems and modify the spatial conditions of urban life, sufficiently abstract as to attract the social imagination. Urban environmentalism was the symbolic construction of collective well-being and common interest through the aestheticization of nature.

Violence was the main reference for urban environmentalism in Medellin, but not the only one. Although inseparable from the illegal drugs trade, violence flared in conditions of economic restructuring, welfare state reform, the privatizing and individualizing tendencies of neoliberalism and growing urban social inequality and poverty. This set of conditions made it increasingly difficult to locate a tenable centre of common interest in the city, yet urban conflict made its reconstruction a matter of urgency. Urban conflict and the search for peace became the political priority and the environment could be constructed as a plausible – though contested – spatial canvas on which such a project could unravel. The transformed natural environment was discursively edified as a unifying dimension of the fragmenting social conditions of existence. Natural elements could be demonstrated as not only flowing under, over and through socio-spatial differentiation, but also as universally affecting the biological existence of all inhabitants and their social quality of life, as well as underpinning the economic and social development of the city as a whole in the new global context.

This was not a consistent policy strategy. Successive urban administration varied enormously in the political importance given to the environment and no mayor could be described as an environmentalist in any strict sense of the word. Environmental programmes were subject to interruptions and changes of emphasis, conflict between institutions was a common feature and meanings of the environment were permanently challenged by social organizations and groups. In other words, urban environmentalism in Medellin was not held together by some broad political agreement and the rational implementation of long-term policy objectives. We argue that it was the rules of the discourse formation on the environment which established its thematic permanence and changes of nuance, and the conditions of emergence of those rules were the crisis of violence and the need to maintain control over a city on the edge of implosion into chaos.

Furthermore, it would be inexact to assume a causal relation between environmental discourse and the defusing of urban violence, or that other areas of transformation in urban social life can directly explain more subtle modifications of environmental meanings. It would be more accurate to view this as a process of general correspondence between discourse and other moments in urban transformation, involving inconsistencies and slippage within a wider dynamic centred around the 'effects of truth' about the environment. The major challenge facing the city authorities was that of maintaining social order, state legitimacy and the moral foundations for the functioning of the urban body. More than a question of power and private interests, the urban challenge was defined in terms of the reconstruction of peaceful coexistence and the collective interest, requiring renewed values, social sensibilities and spatial symbols. This is akin to Foucault's notion of governmentality or the construction of subjectivities and a set of rules for the 'conduct of conduct', thereby contributing to the legitimation of the local state and local government within the changing socio-spatial order of neoliberalism.

The new millennium will be a testing time for urban environmentalism in Medellin. Central government is deconstructing environmental discourse, changing the institutions which mobilize its meanings, imposing new orders of social regulation based on authority, criminal law and the military exercise of power, partly taken up by the local administration. Within it, competitiveness is emerging as the main vector of urban policy, a discourse which transfers survival from the ecological to the economic sphere, privatizes the environment and challenges its integrative function. Nonetheless, the accumulated meanings of the environment are still strong in the city, although the discursive mobilization is now largely in the hands of social organizations and citizen sensibilities.

Chapter 8
Conclusions

The general proposition of this book has been to insist on the need to understand urban environmentalism in the context of the evolving geography of capitalism, and in particular examine the way in which the environment has been constructed so as to contribute to the re-establishment of a sense of urban well-being during a period of radical change and upheaval. Urban environmentalism has been a constitutive part of the major urban transformations of the past two decades or so, character-ized by the conflict inherent in the imposition of the new abstract space of global capital over the lived space of cities. As a consequence, the principal arena of conflict with which we have been concerned is not a general one with nature and natural resource systems – although this of course exists – nor to do with conflicts arising between nation states and regional blocks for the control of the world's resources, although this is also important. The sphere of conflict we have privileged is that of space, and the environment as a means of controlling urban spatial restructuring and mediating the social conflicts which arise within it.

The orthodox view of sustainable development discourse is that environmental improvement is fundamental in making cities competitive and improving the quality of life of urban citizens. As vital nodes in a globalized economy, cities are held to be existing or potential centres of innovation, harmony and progress amid an unstable and perilous international order, and sites where real achievements can be made towards global sustainable development. Cities have undergone enormous transformations, some in a spectacular sense, and almost all can boast one or two showpieces of urban regeneration and new images. The business executive and uncritical tourist can plausibly see cities as greener, cleaner and more enjoyable versions than the often dour and certainly less flamboyant industrial predecessors.

However, now into the new millennium it is easy to forget, especially in the West, the urban struggles and social conflict of the end of the past century. The occasional brilliance of new urban spaces obscures the painful and socially divisive processes of urban economic restructuring. The visible achievements of cities easily obliterate the memory of class and racial conflict, social divides and community tensions that dominated much of the 1980s and 1990s. In most cities in the devel-oped world the expressly political conflict, usually class-based but also arising from the widespread disruption of communities and individual lives, has largely subsided as citizens have adjusted to the demands of a new régime of accumulation. Despite increasing inequality, the descent into poverty for significant sections of the popu-lation has stabilized or been reverted. However, for cities in the developing and former Soviet bloc countries, both poverty and inequality continue to increase. Whilst citizens in the West struggle with lifestyles, those in the developing countries wrestle with livelihoods; in both cases, survival has become an increasingly competitive, exacting and individualized affair.

Neoliberalism, we have argued, is a useful way of describing the regulatory strategy of these changes which, though more usually associated with economic policy, have been exercised most effectively at the city level. Whilst national governments have fashioned macro-economic policy in favour of private enterprise and the spatial displacement of transnational corporations across national boundaries, cities have been given the task of managing its implications for the individual firm, ordinary citizens and everyday lives. Entrepreneurial government, the limitless expansion of the principal of competition and the massive displacement of responsibilities from the state to the individual citizen have reshaped people's urban experience. The city has been reified as the cause of and solution to problems of economic well-being, so that both urban space and citizen sensibilities have been remodelled according to the demands of international competition. This double restructuring means that urban conflict can now take many forms, not just political confrontation and institutionalized conflict but also a whole array of psychological effects, the undermining of collective identities and a sense of individual belonging, violence and criminality, anti-social behaviour, gang warfare, the territorial control of illegal economies, xenophobia, social pathologies, and so on. In short, the restructuring of cities has brought with it fundamental new challenges to the question of urban sociality and public order. In response, extensive and often worrying technological responses have been developed with regard to policing, control and surveillance. 'Living together' has never been more difficult.

We have argued that this is the fundamental issue which urban environmentalism addresses. It is a practical and political issue faced by urban governments, whose legitimacy depends on establishing a sense of order, unity, collective identity and well-being. Seen from this perspective, arguments in favour of 'green governance' appear as an idealized notion of environmental theory; 'green governmentality', with its focus on the power and control, is a much more realistic version of urban events. Urban environmentalism, we hold, is directed towards the problem of the regulation of populations in the postmodern order of global capitalism. In its deregulatory phase, neoliberalism was often brutal and sometimes repressive, but now more often it relies on a subtly coercive modification of citizen subjectivities and people's everyday routines, responsibilities, lifestyles, sense of themselves and their self-worth within changing urban society.

Urban environmentalism fits into this latter category and in its widest sense of 'green governmentality' is a strategic and very urban response to the question of how to regulate social relations, since environment-as-space provides the direct experiential framework of collective existence. As such, the centre of attention is shifted decisively away from the laws of ecology and the rationality of the natural sciences, and in turn the politics of the environment is transferred onto a plane beyond policy issues, institutions and the ecological modernization thesis as it is generally understood in the sense of control over the meanings and resources of sustainable development. The state of the ecological environment and the assigna- tion of blame, the search for explanation or causality – the terrain of conventional environmental politics – cease to be the principle area of concern. Orthodox environmental management is based on the 'externalization' of the environment, the positivist rationality of the environmental sciences and resource management, but in effect it is 'human resources' more than natural resources that are being managed. The requirement of neoliberal urban government is not that of protecting local resource systems but of regulating populations and the creation of docile and useful citizen-consumers. Whether real improvements to the ecological environment are achieved or not is a secondary issue. Whether the social claims of urban environ- mental agendas concerning participatory democracy, improved health, renewed communities and better personal lifestyles, and so on, have any real correspondence to urban life is a matter of little importance. What does matter is the production and reproduction of the 'effects of truth' of urban environmentalism, through the permanent renewal of discourse and spatial form.

In exploring how these 'effects of truth' are achieved, we highlighted the importance of discourse in the creation of meanings around the environment but rejected the temptation to limit analysis to discourse itself. Discourse is 'language in action' which demands looking not just at discourse but at the action as well. Action is partly institutional in the sense of the mobilization of meaning and its translation into concrete interventions in urban space and social life. More import- antly, the action is going on all around, in all the dimensions of urban life or 'moments' of social change. Discourse does not create meaning freely but does so in relation to the actions and interests which those meanings serve.

In a general sense we have described the set of actions on the environment in terms of the aim of neoliberal urban government to accelerate the restructuring of space and the re-regulation of urban populations. Neoliberal governments, by definition, no longer have the instruments of the state as a main strategy, and the environment is constructed as an alternative means through which to mould sensi- bilities, reconstruct notions of citizenship, reorganize values and regulate behaviour in a way which coincides with the demands of new urban economies and the legitimation of public power and authority. In place of public institutions and state authority, urban regulation through the environment is achieved through metaphors and images which connect openly to multiple spheres of social life such as work, consumption, moral responsibility, social status, access to welfare provi- sion and an infinite range of lifestyle issues. Urban environmentalism can therefore be seen as both symbolic and practical, concerned with meanings and effects and, in a general sense, structurally articulated to the forces of globalization which underlie the spatial transformation of cities.

However, in the case of cities and space, urban environmental change is simul- taneously restricted by and potentiated by pre-existing spatial orders. Cities have their own histories, cultural traditions, social conjunctures, political trajectories and institutional configurations, as well as a unique place in the intersecting scales and networks of government. This place specificity determines how, in a broad sense, a city 'talks about itself' and how it receives and re-elaborates international sustainability discourse and urban environmental agendas. In this process, an apparently neutral and technically substantiated environment is ascribed particular meanings and a political purpose in the practice of urban government.

Urban history, which competitive city planning and place promotion is so keen to simplify and exploit as a marketable phenomenon, is precisely that which gives urban environmentalism its indeterminism and political complexity. There is only a general correspondence of urban environmental policy with its practical effects. In the complex process of the spatialization of orthodox urban sustainability policy, environmental discourse encounters all sorts of interferences from other place-specific discourses. 'Slippage' occurs across discursive domains and natural space reacquires its socially-produced character, so that urban environmental management does not result in a uniform mimesis of an abstract 'sustainable city' but produces a locally-determined metamorphosis of a political kind with, in principle at least, open-ended and unpredictable urban consequences.

Exploring how this happens and with what effects was the aim of the city studies. It was illustrated how ecological policy concerning the global responsibility of cities and their performance standards with regard to the state of natural resource systems plays only a supportive role, legitimating the actions of cities on and in the name of the environment without determining the type or significance of such action. In this latter sense the combined demands of economic transformation and local state legitimacy were the driving forces, modulated by political trajectories, social conjunctures, institutional configurations and cultural traditions.

The case of Birmingham, typical of many de-industrializing cities in the West, urban policy was strategically designed to move the city's economy into the financial sector, professional services and the culture industry, and re-imaging the city was a fundamental requirement. As a heavily industrialized city faced with physical obsolescence, this required the construction of environmental meanings around the built environment. However, these meanings were not neutral in their implications, and were necessarily constructed in the interests of the property market which would initiate such urban redevelopment. The environment was the basis of property-led private sector urban regeneration based initially on the city centre, but implicit in this strategy was the need not only to environmentally restructure the urban economic base but also to environmentally refit the labour force. Competitiveness and quality based on the environment had to be extended across the whole of urban space and through all sectors of urban society, including those marginalized and disadvantaged by the new economy. The institutional mobilization of meanings thereby became intimately related to social policy and work-to-welfare programmes. In the case of Birmingham, a tradition in manufacturing and more recent ethnic diversity combined to marginalize the incidence of cultural understandings of nature and environment.

Cities in the formier Soviet bloc faced a similar challenge of economic restructuring but from a radically different historical position. System change from communism to a market economy and liberal democracy involved a new set of epistemological and ideological understandings of nature and the environment, and a radical redistribution of responsibilities between the state, industry and civil society. The mobilization of scientific knowledge was important in bringing about transition but, after communism, urban environmental discourse was replaced by another meta-discourse (now sustainability) imposed by other external institutions (this time the European Union). The imposition of successive environmental dogmas partly explains the democratic deficit and participatory deception of environmentalism in Lodz. Local institutions were initially weak, oriented towards business and susceptible to abuse, and encouraged the rise of a radicalized anti-capitalist fringe. For most citizens, the environment remained a private issue, either as a new opportunity for personal consumption or as an atavistic field of tradition and countryside-based cultural understandings. The results in spatial terms were a privatized, fractured and contradictory urban environment.

Social problems, rather than economic restructuring or political change, determined the construction of environmental meaning in Medellin. Violence of epidemic proportions was discursively mediated through an 'encounter' with abstract nature as a sacred, life-giving and socially-encompassing entity. Human life in general could be cherished and nurtured through the environment. This was made possible by two urban conditions: the aesthetic power of nature in a mountainous tropical climate, and the practical potency of natural forces in everyday urban life as experienced through risk and disasters. Additionally, since violence was concentrated in the popular sectors of the city and related to inequality, then urban environmentalism acquired a socially progressive character. This essentially symbolic construction of the environment meant that scientific knowledge and orthodox urban environmental agendas were of little significance, and the institutional mobilization of meaning was carried out through non-technical organizations using natural resource systems simply as the spatial opportunities for urban interventions. Urban environmentalism in Medellin was a very local and creative construction, eventually susceptible to an encroaching normative (legal and technical) discourse and ideological change at the national level with regard to methods of establishing order.

The overall impression is that the hegemonic discourse of sustainable development and its institutionalized politics are eliminating the urban creativity and socially progressive potential of urban environmentalism. Environmental issues are undoubtedly important in the lives of urban citizens, but there is little convincing empirical evidence to suggest that a decade and a half of sustainable city policy and systematic urban environmental management has improved cities, urban lives or the prospects for the earth's biosphere. Is the whole thing a hopelessly lost cause, with nothing of significance able to be achieved beyond the unwilling or unwitting capitulation to the interests of the restructuring of capital as it affects cities? The generally structuralist approach developed in this book – from the point of view of both political economy and discourse analysis – does indeed tend to suggest that there is not much that can be done, that we are trapped in processes of material change, discursive formations and governmental strategies which frustrate the transformative potential of an environmental perspective at every turn: in urban policy formation, in the city politics, in professional practice and in our personal lives.

It was not the intention of this book to provide answers to these sorts of questions. Nevertheless, the issues raised do have obvious implications for environmentally motivated urban experts and citizens who still place hope of better urban futures in the notion of sustainable cities, and here a recent incident in Medellin is illustrative. A green space in a central neighbourhood was threatened by development. For 30 years this open space had been half cared for and occasionally used, and was generally understood to 'belong' to the housing estate. However, property rights were complex and in the latest land use plan it unexpectedly appeared as a developable site. Local residents immediately organized a civic movement to oppose the development and used all manner of legal devices, publicity and political pressure to stop the proposed new housing development. Apparently, the developers had all permissions in order; the only outstanding requirement was for the felling of some of the trees, which required authorization by the environment authority. Development hinged on this permission, and in a public hearing some sixty interventions were made in defence of the site by an extraordinarily diverse range of social actors all appealing to variations around the argument that protecting the 'environment' and urban ecology were vital to the quality of life, not just for the immediate neighbourhood but for the city as a whole. At the time of writing the outcome was still in the balance and swinging towards the protesters, but the point to be made concerns the potency of the environment to mobilize social groups, the power of environmental discourse to condition and counteract other (legal and administrative) discourses, and frame the defence of urban quality against rapacious developers and untransparent administrative practices.

Behind this minor environmental protest was a sense of injustice and a small rebellion against economic power and careless authority. As people's lives are compacted into stressful routines of survival in an individualized world, nature seems to emerge naturally as something to cling on to, something collective and altruistic in the solitary competitiveness of cities. This is, of course, precisely the effect (of truth) that the discursive construction of the environment is designed to achieve whilst disguising material interests: its ideological function of displacing social contradictions and spatial conflict on to the 'natural' environment. But ideologies can be contested in their own terrain and through urban practice, by demanding that 'effects of truth' become concrete realities. The environment as lived space is strictly regulated and carefully vigilated; unleashing its potential depends not on protecting nature but political renewal.

References

Aberley, D. (1994) *Futures by Design: the practice of ecological planning*, Gabriola Island: New Society.

Acselrad, H. (1999) 'Sustentabilidad y ciudad', *Eure*, 25 (74): 35–46.

Adams, J. (1995) *Risks*, London: UCL Press.

Agnew, J. (1996) *Political Geography, a Reader*, London: Arnold.

Allmendinger, P. and Thomas, H. (eds) (1998) *Urban Planning and the British New Right*, London: Routledge.

Althusser, L. (1969) *For Marx*, London: Harmondsworth Penguin.

Amin, A. (ed.) (1994) *Post-Fordism: a reader*, Oxford: Blackwell.

Amin, A. and Thrift, N. (2002) *Cities: reimagining the urban*, Cambridge: Polity.

Arango, S. (1980) 'La naturaleza desde lo urbano: Bogotá y la generación republicana', *Revista*, 11: 10–18.

Atkinson, R. and Moon, G. (1994) *Urban Policy in Britain: the city, the state and the market*, Basingstoke: Macmillan.

Aydin, Z. (1994) 'Local Agenda 21 and the United Nations', in *First Steps: Local Agenda 21 in Practice*, report from the International Local Authority Local Agenda 21 Conference, 27 June–1 July, Manchester, London: HMSO.

Banister, D. (ed.) (2000) *Transport Policy and the Environment*, London: Spon.

Barber, A. (2001) The ICC, *Birmingham: a catalyst for urban renaissance*, Birmingham: Centre for Urban and Regional Studies, University of Birmingham.

Barton, H. (1998) 'Eco-neighbourhoods: a review of projects', *Local Environment*, 3(2): 159–77.

—— (2000) *Sustainable Communities: the potential for eco-neighbourhoods*, London: Earthscan.

Barton, H. and Tsourou, C. (2000) *Healthy Urban Planning*, London: Spon.

Baudrillard, J. (1993) *The Transparency of Evil: essays on extreme phenomena*, London: Verso.

Bauman, Z. (1992a) *Intimations of Postmodernity*, London: Verso.

—— (1992b) 'Survival as a social construct', *Theory, Culture & Society*, 9: 1–36.

—— (1998) *Work, Consumerism and the New Poor*, Buckingham: Open University Press

—— (2001) *The Individualized Society*, Cambridge: Polity.

—— (2002) 'Individually, together', foreword to U. Beck and E. Beck-Gernsheim, *Individualization*, London: Sage.

Bauriedl, S. and Wissen, M. (2002) 'Post-Fordist transformation, the sustainability concept and social relations with nature: a case study of the Hamburg region', *Journal of Environmental Policy and Planning*, 4: 107–21.

Beazley, M. and Loftman, P. (2001) *Race and regeneration: black and minority ethnic experience of the Single Regeneration Budget*, London: Local Government Information Unit/London Borough of Camden.

Beck, U. (1992a) 'From industrial society to risk society: questions of survival, social structure and ecological enlightenment', *Theory Culture & Society*, 9:97–123.

Beck, U. (1992b) *Risk Society: towards a new modernity*, London: Sage.
—— (1996) 'World risk society as cosmopolitan society? Ecological questions in a framework of manufactured uncertainty', *Theory, Culture and Society*, 13(4): 1–32.
Beck, U. and Beck-Gernsheim, E. (2002) *Individualization*, London: Sage.
Beck, U., Giddens, A. and Lash, S. (1994) *Reflexive Modernization*, Cambridge: Polity.
Bejarano, A. M. (1992) 'Democracia y sociedad civil', *Análisis Político*, 15: 97–104.
Bentley I., Alcock A., Murrain P., McGlynn S., and Smith G. (1985) *Responsive Environments*, Oxford: Butterworth-Heinemann.
Betancur, J. (2004) 'Medellin y la cultura del rebusque', in P. Navia and M. Zimmerman (eds) *Las ciudades latinoamericanas en el nuevo (des)orden mundial*, Mexico: Siglo XXI.
Birmingham City Council (1987) *City Centre Strategy*, Department of Architecture and Planning, Birmingham.
—— (1989) *The Highbury Initiative, Report of Proceedings of the City Challenge Symposium*, Birmingham.
—— (1992a) *Third City Centre Review*, Department of Planning and Architecture, Birmingham.
—— (1992b) *Conservation Strategy for Birmingham*, Department of Planning and Architecture, Birmingham.
—— (1993) *The Birmingham Plan: Birmingham Unitary Development Plan 1993*, Department of Planning and Architecture, Birmingham.
—— (1993) *The Green Action Plan*, Birmingham.
—— (1995a) *Moving Forward Together: Birmingham's City Pride Prospectus*, Birmingham.
—— (1995b) *Economic Development Strategy for Birmingham, 1995–1998*, Economic Development Department, Birmingham.
—— (1997) *Living Today with Tomorrow in Mind*, Birmingham.
—— (1998) *Summit of the Cities: Developing Our Common Agenda*, Birmingham.
—— (2000a) *Sustainability Strategy and Action Plan 2000–2005*, Birmingham.
—— (2000b) *Review and Assessment of Air Quality in Birmingham*, www.birmingham.gov. uk/environment (consulted 20 March 2004).
—— (2002) *Local Agenda 21, Achievements 1992–2002*, www.birmingham.gov.uk/environment/ sustainability (consulted 18 March 2004).
—— (2003) *The Birmingham Economy*, www.birminghameconomy.org.uk (consulted 10 September 2003).
—— (2004a) *Sustainability achievements*, www.birmingham.gov.uk/environment/sustainabil- ity (consulted 20 March 2004).
—— (2004b) *Regeneration*, www.birmingham.gov.uk/economy/regeneration (consulted 18 March 2004).
Blackman, T. (1995) *Urban Policy in Practice*, London: Routledge.
Blowers, A. (ed.) (1993) *Planning for a Sustainable Environment*, London: TCPA/Earthscan.
—— (2000) 'Britain: unsustainable cities', in N. Low, B. Gleeson, I. Elander and R. Lidskog (eds) *Consuming Cities*, London: Routledge.
Bond, P. (2002) 'Reportback: A Johannesburg roundtable debate on "sustainable development" and fixing/nixing the W$$D', *Capitalism, Nature, Socialism*, 13(4): 81–4.
Bordessa, R. (1993) 'Geography, postmodernism, and environmental concern', *The Canadian Geographer*, 37(2): 147–56.
Bourdieu, P. (1998) *Acts of Resistance: against the tyranny of the market*, New York: Free Press.
Brand, P. (1996) 'Urban environmentalism: in the twilight between vice and virtue, *Urban Design International*, 1(4): 357–60.
—— (2000) 'The sustainable city as metaphor: urban environmentalism in Medellin, Colombia', in M. Jenks and R. Burgess (eds) *Compact Cities: sustainable urban forms for developing countries*, London: Spon.

—— (ed.) (2001a) *Trayectorias Urbanas en la Modernización del Estado Local en Colombia*, Bogota: Tercer Mundo.

—— (2001b) 'La construcción ambiental del bienestar urbano: el caso de Medellín, Colombia', *Economía, Sociedad y Territorio*, 3(9): 1–24.

Brand, P. and Prada, F. (2003) *La Invención de Futuros Urbanos: competitividad económica y sostenibilidad ambiental en las cuatro ciudades principales de Colombia*, Medellín: Universidad Nacional de Colombia.

Breheny, M. (1992) 'The contradictions of the compact city', in M. Breheny (ed.) *Sustainable Development and Urban Form*, London: Pion.

Brenner, N. and Theodore, N. (eds) (2002) *Spaces of Neoliberalism*, Oxford: Blackwell.

Briggs, A. (1968) *Victorian Cities*, London: Harmondsworth Penguin.

Bryant, R. L. (1997) 'Beyond the impasse: the power of political ecology in Third World environmental research', *Area*, 29(1): 5–19.

Burgess, R., Carmona, M. and Kolstee, T. (eds) (1997) *The Challenge of Sustainable Cities: neoliberalism and urban strategies in developing countries*, London: Zed Books.

Bustamante, M. (1987) '*Los desastres "naturales" en Medellín*' (masters thesis), School of Urban and Regional Planning, Universidad Nacional de Colombia, Medellín.

—— (1988) 'Inventario de desastres recientes de origen geológico en el Valle de Aburrá' (unpublished paper), *II Conferencia de Riesgos Geológicos en el Valle de Aburrá*, 2–6 August, Medellin.

Callinicos, A. (1989) *Against Postmodernism: a Marxist critique*, Oxford: Blackwell.

Carley, M. Jenkins P., and Smith, H. (eds) (2001) *Urban Development and Civil Society: the role of communities in sustainable cities*, London: Earthscan.

Castells, M. (1983) 'Crisis, planning and the quality of life: managing the new historic relations between space and society', *Environment and Planning D: Space and Society*, 1: 3–21.

—— (1989) *The Informational City*, Oxford: Blackwell.

—— (1996) *The Rise of the Network Society*, Oxford: Blackwell.

—— (1996–98) *The Information Age: Economy, Society and Culture*, 3 vols: *I The Rise of the Network Society; II The Power of Identity; III End of the Millennium*, Oxford, Blackwell.

—— (1997) *The Power of Identity*, Oxford: Blackwell.

Castree, N. (1997) 'The nature of produced nature: materiality and knowledge construction in Marxism', *Antipode*, 27(1): 12–48.

Chapman, D. W. and Larkham, P. J. (1999) 'Urban design, urban quality and the quality of life: reviewing the Department of the Environment's Urban Design Campaign', *Journal of Urban Design*, 4(2): 211–32.

Churski, P. (2002) 'Unemployment and labour-market policy in the new voivodeship system in Poland', *European Planning Studies*, 10(6): 745–63.

Clark, N. (1997) 'Panic ecology: nature in the age of superconductivity', *Theory, Culture & Society*, 4(1): 77–96.

Cole, D. H. (1998) *Instituting Environmental Protection: From Red to Green in Poland*, Basingstoke: Macmillan.

Comeliau, C. (2000) 'The limitless growth assumption', *International Social Science Journal*, 166: 457–66.

Commission of the European Communities (1990) *Green Paper on the Urban Environment*, Brussels: CEC.

Cottle, S. (2000) 'TV news, lay voices and the visualization of environmental risks', in S. Allan, B. Adams and C. Carter (eds) *Environmental Risks and the Media*, London: Routledge.

Cousins, M. and Hussain, A. (1984) *Michel Foucault*, London: Macmillan.

Crean, C. (1998) Personal interview, Birmingham Friends of the Earth.

Czyz, T. (2002) 'Regional inequalities in Poland and the country's new territorial organisation', in R. Domanski (ed.) *Cities and Regions in an Enlarging European Union*, Warsaw: Polish Academy of Sciences.

Daly, H. (1996) *Beyond Growth: the economics of sustainable development*, Boston: Beacon Press.

Damian, M. and Graz, J. C. (2001) 'The World Trade Organisation, the environment and the ecological critique', *International Social Science Journal*, 170: 597–610.

Darrier, E. (ed.) (1999) *Discourses of the Environment*, Oxford: Blackwell.

Department of the Environment (1994) *Quality in Town and Country*, London: DoE.

Department of the Environment, Transport and the Regions (1999) *A Better Quality of Life: Sustainable Development Strategy for the UK*, London: DETR.

Douglas, M. (1992) *Risks and Blame: essays in cultural theory*, London: Routledge.

Dragsbaek Achmidt, J. (1998) 'Globalization and inequality in urban South-east Asia', *Third World Planning Review*, 20(2): 127–45.

Dryzek, J. S., Hunold, C., Schlosberg, D. and Downes, D., Hernes, H-K. (2002) 'Environmental transformation of the state: the USA, Norway, Germany and the UK', *Political Studies*, 50: 659–82.

Dwyer, P. (2000) *Welfare Rights and Responsibilities*, Bristol: Policy Press.

Eagleton, T. (1991) *Ideology: an introduction*, London: Verso.

Edén M., Falkheden L. and Malbert, B. (2000) 'The built environment and sustainable development: research meets practice in a Scandinavian context', *Planning Theory and Practice*, 1(2): 260–72.

El Espectador (1997), Sunday 23 March, p. 5A.

Environment Agency Midlands Region (2004) Homepage, www.environment-agency.gov.uk/regions/midlands (consulted 12 March 2004).

Escobar, A. (1996) 'Constructing nature: elements for a postmodern political ecology', in R. Peet and M. Watts (eds) *Liberation Ecologies: environment, development, social movements*, London: Routledge.

European Commission (2004) *European Common Indicators*, www.europa.eu.int/comm/environment/urban/common_indicators (consulted 10 March 2004).

European Foundation for the Improvement of Living and Working Conditions (1996) *Utopias and Realities of Urban Sustainable Development*, Conference Proceedings, Turin-Barolo, 19–21 September, Turin.

Evans, B. and Rydin, Y. (1997) 'Planning, professionalism and sustainability', in A. Blowers and B. Evans (eds) *Town Planning into the 21st Century*, London: Routledge.

Fairclough, N. (1992) *Discourse and Social Change*, Cambridge: Polity.

Febres Cordero, J. B. (1991) *El Proceso Constituyente*, Bogota: Camara de Representantes/ Universidad Pontificia Javeriana.

Ferry, M. (2002) 'The Polish green movement ten years after the fall of communism', *Environmental Politics*, 11(1): 172–7.

Fischer, F. (2000) Citizens, *Experts and the Environment: the politics of local knowledge*, Durham: Duke University Press.

Fischer, F. and Hajer, M. (eds) (1999) *Living with Nature: environmental politics as cultural discourses*, Oxford: Oxford University Press.

Fitzpatrick, M. (1995) 'An epidemic of fear', *Living Marxism*, No. 84.

Flyvbjerg, B. (1998) *Rationality and Power: democracy in practice*, Chicago: Chicago University Press.

—— (2001) *Making Social Science Matter*, Cambridge: Cambridge University Press.

Forester, J. (1993) *Critical Theory, Public Policy and Planning Practice*, Albany: State University of New York Press.

Foucault, M. (1979) *Discipline and Punish: the birth of the prison*, Harmondsworth: Penguin.
—— (1980) *Power/Knowledge: selected interviews and other writings 1972–1977*, New York: Pantheon Books.
—— (1981) *The History of Sexuality 1: the will to knowledge*, Harmondsworth: Penguin.
—— (1989) *The Archaeology of Knowledge*, London: Routledge.
—— (1991) 'Governmentality', in G. Burchell, C. Gordon and P. Miller (eds) (1991) *The Foucault Effect: Studies in Governmentality*, Chicago: University of Chicago Press.
Franklin, J. (1998) *The Politics of Risk Society*, Cambridge: Polity.
Friends of the Earth (2003) *Practical Steps for a Greener Birmingham*, www.birminghamfoe.org.uk (consulted 12 March 2004).
Frolov, I. (1981) 'La concepción Marxista–Leninista acerca del problema ecológico', in *La Sociedad y el Medio Ambiente: concepción de los científicos soviéticos*, Moscow: Editorial Progreso.
Frow, J. (1985) 'Discourse and power', *Economy and Society*, 14.
Gaceta Constitucional (1991) No. 46, 15 April 1991, and No. 94, 11 June 1991, Bogota: Presidency of the Republic of Colombia.
Gandy, M. (1996) 'Crumbling land: the postmodernity debate and the analysis of environmental problems', *Progress in Human Geography*, 20(1): 23–40.
Garay, L. J. (2002) *Colombia: entre la exclusión y el desarrollo*, Bogota: Contraloria General de la Nacion.
Gare, A. (1995) *Postmodernism and the Environmental Crisis*, London: Routledge.
Giddens, A. (1990) *The Consequences of Modernity*, Cambridge: Polity.
—— (1994) *Beyond Left and Right*, Cambridge: Polity.
Gilbert, R., Stevenson, D., Girardet, H. and Stren, R. (1996) *Making Cities Work: the role of local authorities in the urban environment*, London: Earthscan.
Girardet, H. (1992) *The Gaia Atlas of Cities: new directions for sustainable urban living*, London: Gaia Books.
Gleeson, B. and Low, N. (2000) 'Cities as consumers of the world's environment', in N. Low, B. Gleeson, I. Elander, and R. Lidskog (eds) (2000) *Consuming Cities*, London: Routledge.
Goldman, M. (2002) 'Notes from the World Summit in Johannesburg: "History in the making?"', *Capitalism, Nature, Socialism*, 13(4): 68–79.
Goodchild, B. and Cole, I. (2001) 'Social balance and mixed neighbourhoods in Britain since 1979: a review of discourse and practice in social housing', *Environment and Planning D: Society and Space*, 19: 103–21.
Gorz, A. (1978) *Ecology as Politics*, Boston: South End Press.
Graham, S. and Marvin, S. (2001) *Splintering Urbanism*, London: Routledge.
Gram-Hanssen, K. (2000) 'Local Agenda 21: traditional *gemeinschaft* or late-modern subpolitics?', *Journal of Environmental Policy and Planning*, 2: 225–35.
Greco, M. (1993) 'Psychosomatic subjects and the "duty to be well": personal agency within medical rationality', *Economy and Society*, 22(3): 357–71.
Green, J. (1997) *Risk and Misfortune: a social construction of accidents*, London: UCL Press.
Green, N. (1990) *The Spectacle of Nature: landscape and bourgeois culture in nineteenth-century France*, Manchester: Manchester University Press.
Griffiths, R. (1998) 'Making sameness: place marketing and the new urban entrepreneurialism', in N. Oatley (ed.) *Cities, Economic Competition and Urban Policy*, London: Chapman.
Grove-White, R. (1992) 'Environmentalism: a new moral discourse for technological society?', in K. Milton (ed.) *Environmentalism: a view from anthropology*, London: Routledge.
Guhl, E. (1992) 'Comentarios al plan de desarrollo económico y social 1990–1994', in E. Guhl and J.Tokatlian (eds) *Medio Ambiente y Relaciones Internacionales*, Bogota: Tercer Mundo.

Hajer, M. (1995) *The Politics of Environmental Discourse: ecological modernization and the policy process*, Oxford: Clarendon Press.

Hajer, M. and Kesselring, S. (1999) 'Democracy in the risk society? Learning from the new politics of mobility in Munich', *Environmental Politics*, 8 (3): 1–23.

Hall, T. and Hubbard, P. (1998) *The Entrepreneurial City: geographies of politics, régime and representation*, Chichester: Wiley.

Hall, P. and Pfeiffer, U. (2000) Urban Future 21: *A global agenda for twenty-first century cities*, London: Spon.

Hannigan, J. (1995) *Environmental Sociology: a social constructionist perspective*, London: Routledge.

Haraway, D. (1991) *Simians, Cyborgs and Women: the reinvention of nature*, London: Free Association Books.

Hardoy, J, Mitlin D. and Satterthwaite, D. (2001) *Environmental Problems in an Urbanizing World*, London: Earthscan.

Harré, R., Brockmeier J. and Mühlhäusler P. (1999) *Greenspeak: A study of environmental discourse*, London: Sage.

Harvey, D. (1982) *The Limits to Capital*, Oxford: Blackwell.

—— (1989) *The Urban Experience*, Oxford: Blackwell.

—— (1990) *The Condition of Postmodernity: an enquiry into the origins of cultural change*, Oxford: Blackwell.

—— (1996) *Justice, Nature and the Geography of Difference*, Oxford: Blackwell.

—— (2000) *Spaces of Hope*, Edinburgh: Edinburgh University Press.

Haughton, G. (1999) 'Environmental justice and the sustainable city', in D. Satterthwaite (ed.) *Sustainable Cities*, London: Earthscan.

Haughton G. and Hunter C. (1994) *Sustainable Cities*, London: Jessica Kingsley/Regional Studies Association.

Hayward, R. and McGlynn S. (eds) (1993) *Making Better Places*, Oxford: Butterworth-Heinemann.

Healey, P. (1996) 'Planning through debate: the communicative turn in planning theory', in S. Fainstein and S. Campbell (eds) *Readings in Planning Theory*, Oxford: Blackwell.

Heller, A. (1988) *The Postmodern Political Condition*, Cambridge: Polity.

Hewitt, M. (2001) 'New labour, human nature and welfare reform', in R. Sykes, C. Bochel and N. Ellison (eds) *Social Policy Review 13, Developments and Debates 2000–2001*, Bristol: Policy Press.

Holmberg, J., Thompson, K. and Timberlake, L. (1993) *Facing the Future: beyond the Earth Summit*, London: IIED/Earthscan.

Holmes, L. (1997) *Post-Communism: an introduction*, Cambridge: Polity.

Hoogma, R., Kemp, R., Schot, J. and Truffer, B. (2002) *Experimenting for Sustainable Transport*, London: Spon.

Hough, M. (1995) *Cities and Natural Process*, London: Spon.

Howarth, D. (2002) 'An archaeology of political discourse? Evaluating Michel Foucault's explanation and critique of ideology' (and reply by S. Mulligan, in same issue), *Political Studies*, 50: 117–35.

Høyer, K. G. and Næss, P. (2001) 'The ecological traces of growth: economic growth, liberalization, increased consumption – and sustainable urban development?', *Journal of Environmental Policy and Planning*, 3: 177–92.

Hoyos, G. (1989) 'Elementos para una ética ambiental', in *Ciencias Sociales y Medio Ambiente*, Bogota: ICFES.

Huxley, M. (2002) 'Governmentality, gender, planning', in P. Allmendinger and M. Tewdwr-Jones (eds) *New Directions for Planning Theory*, London: Routledge.

Huyssens, A. (1984) 'Mapping the postmodern', *New German Review*, 33.

Isin, E. H. (1998) 'Governing Toronto without government: liberalism and neoliberalism', *Studies in Political Economy*, 56: 169–91.

Italian Environmental Forum (2002) 'Sustainable development or the law of profit', *Capitalism, Nature, Socialism*, 13(4): 63–7.

Jameson, F. (1984) 'Postmodernism, or the cultural logic of late capitalism', *New Left Review*, 146: 53–92.

Jaramillo, A. M. (1995) 'Control social y criminalidad en el Medellin del siglo XX', *Desde la Region*, 19: 22–4.

—— (1996) 'No era culpa de Pablo Escobar: Medellin sigue entre la vida y la muerte', *Desde la Region*, 21: 12–15.

Jenks, M., Burton, E. and Williams, K. (eds) (1996) *Compact Cities, a Sustainable Urban Form?*, London: Spon.

Jenks, M. and Burgess R. (eds) (2000) *Compact Cities: sustainable urban forms for developing countries*, London: Spon.

Jessop, B. (1997) 'The entrepreneurial city: re-imaging localities, re-designing economic governance or restructuring capital?' in N. Jewson and S. MacGregor (eds) *Transforming Cities*, London: Routledge.

—— (2001) 'The crisis of the national spatio-temporal fix and the ecological dominance of globalizing capitalism', *International Journal of Urban and Regional Studies*, 24: 273–310.

—— (2002) 'Liberalism, neoliberalism and urban governance: a state-theoretical perspective', in N. Brenner and N. Theodore (eds) *Spaces of Neoliberalism*, Oxford: Blackwell.

Jones, M. and Ward, K. (2002) 'Excavating the logic of British urban policy: neoliberalism and the "crisis of crisis management"', in N. Brenner and N. Theodore (eds) *Spaces of Neoliberalism*, Oxford: Blackwell.

Kaczmarek, S. (1997) 'Spatial differentiation of housing conditions and urban landscape in Lodz', in S. Liszewski and C. Young (eds) *A Comparative Study of Lodz and Manchester: geographies of european cities in transition*, Lodz: University of Lodz.

Keil, R. (2002) '"Common-sense" neoliberalism: progressive Conservative neoliberalism in Toronto, Canada', in N. Brenner and N. Theodore (eds) *Spaces of Neoliberalism*, Oxford: Blackwell.

Klasik, A. (2002) 'Regional strategies: Polish experiences', in R. Domanski (ed.) *Cities and Regions in an Enlarging European Union*, Warsaw: Polish Academy of Sciences.

Kozakiewicz, M. (1996) *The analysis of the current situation of Environmental NGOs in Poland*, Working Paper, Lodz: University of Lodz.

Kurczewski, J. and Kurczewska, J. (2001) 'A self-governing society twenty years after: democracy and the third sector in Poland', *Social Research*, 68(4): 937–76.

Lanthier, I. and Olivier, L. (1999) 'The construction of environmental awareness', in E. Darrier (ed.) *Discourses of the Environment*, Oxford: Blackwell.

Lash, S. (ed.) (1996) *Risk, Environment, Modernity: towards a new ecology*, London: Sage.

—— (2002) 'Individualization in a non-linear mode', foreword to U. Beck and E. Beck-Gernsheim, *Individualization*, London: Sage.

Lechner, N. (1989) 'Ese desencanto llamado posmoderno', *Foro*, 10: 35–45.

Lefebvre, H. (1976) *The Survival of Capitalism* (first edition in French in 1973), London: Allison and Busby.

—— (1991) *The Production of Space* (first published in French in 1974), Oxford: Blackwell.

Lein, J. K. (2003) *Integrated Environmental Planning*, Oxford: Blackwell.

Lipietz, A. (1996) 'Geography, ecology, democracy', *Antipode*, 28(3): 219–28.

Lipovetsky, G. (1990) *L'Ere du Vide*, Paris: Gallimard.

Liszewski, S. (1997) 'The origins and stages of development of industrial Lodz and the Lodz urban region', in S. Liszewski and C. Young (eds) *A Comparative Study of Lodz and Manchester: geographies of european cities in transition*, Lodz: University of Lodz.

Liszewski, S., Kaczmarek, S. and Kaczmarek, J. (1995) 'Geographical studies of a Polish city', *Research Paper 8*, London: Queen Mary and Westfield College, University of London.

Lockie, S. (2000) 'Environmental governance and legitimation: state-community interactions and agricultural land degradation in Australia', *Capitalism, Nature Socialism*, 11(2): 41–58.

Lodz City Office (1997) *Ecological Policy for the City of Lodz*, Lodz.

—— (2001) *Directions of Actions for Health of Lodz citizens*, Lodz.

—— (2002) *Lodz in Figures*, Lodz.

Lofsted, R. (1997) *Earthscan Reader in Risk and Modern Society*, London: Earthscan.

Loftman, P. and Nevin, B. (1992) *Urban Regeneration and Social Equality: a case study of Birmingham 1986–1992*, Faculty of the Built Environment Research Paper No. 8, University of Central England, Birmingham.

—— (1994) 'Prestige project developments: economic renaissance or economic myth? A case study of Birmingham', *Local Economy*, 8(4): 307–25.

—— (1996) 'Going for growth: prestige projects in three British cities', *Urban Studies*, 33(6): 991–1019.

Low, N. (2002) 'Ecosocialisation and environmental planning: a Polanyian approach', *Environment and Planning A*, 36: 43–60.

Low, N., Gleeson B., Elander, I. and Lidskog, R. (eds) (2000) *Consuming Cities: the urban environment in the global economy after the Rio declaration*, London: Routledge.

Luke, T. W. (1997) 'At the end of nature: cyborgs, "humachines", and environments in post-modernity', *Environment and Planning A*, 29: 1367–80.

Lyotard, J.-F. (1984) *The Postmodern Condition*, Manchester: Manchester University Press.

McGranahan, G., Jacobi, P., Songsore J., Surjadi C. and Kjellén M. (2001) *The Citizens at Risk: from urban sanitation to sustainable cities*, London: Earthscan.

McHarg, I. (1969) *Design with Nature*, New York: Natural History Press.

Macnaghten, P. and Urry, J. (1998) *Contested Natures*, London: Sage.

Maffesoli, M. (1991) 'The ethics of aesthetics', *Theory, Culture & Society*, 8: 7–20.

Marcuse, P. (1998) 'Sustainability is not enough', *Environment and Urbanization*, 10(2): 103–11.

Markowski, T. (2002) 'Bipolar urban development: opportunities, threats and sustainability – the case of the Warsaw and Lodz agglomerations', in R. Domanski (ed.) *Cities and Regions in an Enlarging European Union*, Warsaw: Polish Academy of Sciences.

Markowski, T. and Marszal, T. (1999) 'Recovering economy of a region in transition: the case of the Lodz Industrial Agglomeration (Poland)', *European Spatial Research and Policy*, 6(1): 31–52.

Markowski, T. and Rouba, H (2000) 'Poland: on the way to a market economy', in N. Low, B. Gleeson, I. Elander and R. Lidskog (eds) *Consuming Cities*, London: Routledge.

Marshall, T. (1994) 'British planning and the New Environmentalism', *Planning Practice and Research*, 9(1): 21–30.

—— (1996) 'Dimensions of sustainable development and scales of policy-making', in S. Baker, M. Kousis, D. Richardson and S. Young (eds) *The Politics of Sustainable Development: theory, policy and practice within the European Union*, London: Routledge.

Marx, K. (1988) *The Economic and Philosophic Manuscripts of 1844*, New York: Prometheus Books.

—— (1990) *Capital* (Vol. 1), London: Penguin.

Meadows, D. H., Meadows, D. L., Randers, J. and Behrenvs, W. W. (1972) *The Limits to Growth: a report for the Club of Rome's project on the predicament of mankind*, London: Pan.

Medellin City Council (1989) *Problemática ambiental del Valle de Aburra*, Medellin: Municipio de Medellin.

—— (1997) *Balance de Gestión 1995–1996*, Medellin: Instituto Mi Rio.

Medellin Municipality/UNDP (1996) *Medellín, los esfuerzos de una ciudad muestran que la prevención da resultados*, Medellin: UNDP/Municipio de Medellin.

Medellin's Presidential Programme (1992) *Medellín: en el camino a la concertación, Informe de Gestión 1990–1992*, Medellin: Presidencia de la Republica.

Medellin Planning Department (1985) *Plan de Desarrollo Metropolitano del Valle de Aburra*, Medellin: Municipio de Medellin.

—— (1986) *Plan de Desarrollo de Medellín 1986*, Medellin: Municipio de Medellin.

—— (1990) *Plan de Desarrollo de Medellín*, Medellin: Municipio de Medellin.

—— (1993) *Plan General de Desarrollo para Medellín*, Medellin: Municipio de Medellin.

—— (1996) *Plan de Desarrollo de Medellín, 1995–1997*, Medellin: Municipio de Medellin.

—— (1999) *Plan de Ordenamiento Territorial*, (Technical Support Document 1), Medellín: Municipio de Medellín.

—— (2000) *Plan de Desarrollo 'Medellín Competitivo'*, Medellin: Municipio de Medellin.

Mendes, C. (ed.) (1977) El Mito del Desarrollo, Barcelona: Kairos.

Meyer-Bisch, P. (2001) 'Social actors and sovereignty in IGOs', *International Journal of Social Science*, 170: 597–610.

Middleton, N. and O'Keefe, P. (2003) *Rio Plus Ten: politics, poverty and the environment*, London: Pluto Press.

Miller, D. and de Roo, G. (1999) *Integrating City Planning and Environmental Improvement*, Aldershot: Ashgate.

Ministry of Economic Development (1996) *Ciudades y Ciudadanía*, Bogota: Ministerio de Desarrollo Economico.

Mittler, D. (2001) 'Hijacking sustainability? Planners and the promise and failure of Local Agenda 21', in A. Layard, S. Davoudi and S. Batty (eds) *Planning for a Sustainable Future*, London: Spon.

Moncayo, V. M. (2003) *'La realidad neoliberal'*, in D. Restrepo (ed.) La Falacia Neoliberal, Bogota: Universidad Nacional de Colombia.

Mongardi, C. (1992) 'The ideology of postmodernity', *Theory, Culture & Society*, 9: 55–65.

Moughtin, C. (1996) *Urban Design – Green Dimensions*, Oxford: Butterworth-Heinemann.

Murie, A., Beazley, M. and Carter, D. (2003) 'The Birmingham Case', in W. Salet, A. Thornley and A. Kreukels (eds) *Metropolitan Governance and Spatial Planning: comparative case studies of European city regions*, London: Routledge.

National Office for the Prevention and Attention of Disasters (ONAD) (1991) *Programa de mitigación de desastres en Colombia 1988–91: Informe de un proyecto*, Bogota: ONAD/UNDRO.

Neal, P. (2003) *Urban Villages and the Making of Communities*, London: Spon.

Newman, P. (1999) 'Transport: reducing automobile dependence', in D. Satterthwaite (ed.) *Sustainable Cities*, London: Earthscan.

Newman, P. and Thornley, A. (1996) *Urban Planning in Europe: international competition, national systems and planning projects*, London: Routledge.

Nijkamp, P. and Perrels, A. (1994) *Sustainable Cities in Europe: a comparative analysis of urban energy-environmental policies*, London: Earthscan.

O'Connor, M. (ed.) (1994) *Is Capitalism Sustainable? Political economy and the politics of ecology*, London: Guildford Press.

OECD (1990) *Environmental Policies for Cities in the 1990s*, Paris: OECD.

OECD (1996) *Innovative Policies for Sustainable Urban Development: the ecological city*, Paris: OECD.

Offe, C. (1996) *Varieties of Transition*, Cambridge: Polity Press.

O'Riordan, T. (1995) *Perceiving Environmental Risks*, London: Academic Press.

Osborne T. and Rose, N. (1999) 'Governing cities: notes on the spatialisation of virtue' *Environment and Planning D: Society and Space*, 17: 737–60.

O'Tuathail, G. (1998) 'Political geography III: dealing with deterritorialisation', *Progress in Human Geography*, 22(1): 81–93.

Palacio, G. (1994) 'Notas preliminares sobre la redefinición jurídica de las relaciones sociales con la naturaleza', in *Derecho y Medio Ambiente* (Proceedings of the II National Seminar on Environmental Law), Medellin: Penca de Sábila.

Pearce, D., Markandya, A. and Barbier, E. (1989) *Blueprint for a Green Economy*, London: Earthscan.

Petrella, R. (1996) 'Sustainable cities: towards new alliances for solidarity among the generations and across cities', in *Utopias and Realities* (Conference Proceedings), Turin-Barolo, 19–21 Sept. 1996, European Foundation for the Improvement of Living and Working Conditions.

Pfeil, F. (1988) 'Postmodernism as a structure of feeling', in C. Nelson and L. Grossberg (eds) *Marxism and the Interpretation of Culture*, London: Macmillan.

Plataforma Colombiana de Derechos Humanos, Democracia y Desarrollo (2003) *El Embrujo Autoritario: primer año del gobierno de Álvaro Uribe Vélez*, Bogota: Plataforma Colombiana de Derechos Humanos, Democracia y Desarrollo.

Polish Ministry of the Environment (2002) *Agenda 21: 10 years After Rio*, Warsaw.

Powell, M. (1999) *New Labour, New Welfare State?*, Bristol: Policy Press.

Pratt, D. (2000) 'Greening the structural funds: the case of an English region 1993–2006', *Local Governance*, 26(4): 247–70.

Pugh, C. (ed.) (1996) *Sustainability, the Environment and Urbanization*, London: Earthscan.

—— (ed.) (2000) *Sustainable Cities in Developing Countries*, London: Earthscan.

Pyszkowski, A. (1998) 'Regional policy – conditions and dilemmas', in R. Domanski (ed.) *Emerging Spatial and Regional Structures of an Economy in Transition*, Warsaw: Polish Academy of Sciences.

Rabinow, P. (ed.) (1984) *The Foucault Reader*, London: Penguin.

Register, R. (2002) *Ecocities: building cities in balance with nature*, Berkeley: Berkeley Hills Books.

Restrepo, D. (ed.) (2003) *La Falacia Neoliberal*, Bogota: Universidad Nacional de Colombia.

Richardson, T. (2002) 'Freedom and control in planning: using discourse in the pursuit of reflexive practice', *Planning Theory and Practice*, 3(3): 353–61.

Roberts, P. and Sykes, H. (eds) (2000) *Urban Regeneration: a handbook*, London: Sage.

Rodriguez Becerra, M. (1992) 'Medio ambiente y desarrollo en la nueva constitución política de Colombia', in E. Guhl and J. Tokatlian (eds) *Medio Ambiente y Relaciones Internacionales*, Bogota: Tercer Mundo.

Roelofs, J. (1996) *Greening Cities: building just and sustainable communities*, New York: Bootstrap Press.

Rogers, R. and Gumuchdjian, P. (1997) *Cities for a Small Planet*, London: Faber and Faber.

Rose, N. (1996) 'Governing "advanced" liberal democracies', in A. Barry, T. Osborne and N. Rose (1996) *Foucault and Political Reason: liberalism, neo-liberalism and rationalities of government*, London: UCL Press.

Roseland, M. (1998) *Towards Sustainable Communities*, Gabriola Island: New Society Publishers.

Royal Society (1992) *Risk: analysis, perception, management*, London: Royal Society.

Rutherford, P. (2003) '"Talking the talk": business discourse at the World Summit on sustainable development', *Environmental Politics*, 12(2): 145–50.

Rydin, S. (2003) *Conflict, Consensus and Rationality in Environmental Planning: an institutional discourse approach*, Oxford: Oxford University Press.

Rydin, Y. and Thornley, A. (eds) (2002) *Planning in the UK: agendas for the new millennium*, Aldershot: Ashgate.

Sachs, W. (1992) 'Environment', in W. Sachs (ed.) *The Development Dictionary*, London: Zed Books.

Sagalara (2002) *Leave Your Heart in Lodz*, Lodz: Oficyna Wydawniczo-Reklamowa Sagalara.

Samuels, R. and Prasad, D. K. (1996) *Global Warming and the Built Environment*, London: Spon.

Sandercock, L. (2003) 'Out of the closet: the importance of stories and storytelling in planning practice', *Planning Theory and Practice*, 4(1): 11–28.

Sassen, S. (2001) *The Global City*, New Jersey: Princeton University Press.

Satterthwaite, D. (ed.) (1999) *Sustainable Cities*, London: Earthscan.

Schmidt, A. (1971) *The Concept of Nature in Marx* (first edition in German in 1962), London: New Left Books.

Sharp, L. and Richardson, T. (2001) 'Reflections on Foucauldian discourse analysis in planning and environmental policy research', *Journal of Environmental Policy and Planning*, 3: 193–209.

Smart, B. (1983) *Foucault, Marxism and Critique*, London: Routledge.

Smith, N. (1984) *Uneven Development: nature, capital and the production of space*, Oxford: Blackwell.

Smith J., Blake J. and Davies A. (2000) 'Putting sustainability in place: sustainable communities projects in Huntingdonshire', *Journal of Environmental Policy and Planning*, 2: 211–23.

Soja, E. W. (1989) *Postmodern Geographies: the reassertion of space in critical social theory*, London: Verso.

—— (1996) *Thirdspace*, Oxford: Blackwell.

—— (2000) *Postmetropolis: critical studies of cities and regions*, Oxford: Blackwell.

Starzewska, A. (1987) 'The Polish People's Republic', in G. Enyedi *et al. Environmental Policies in East and West*, London: Graham Taylor.

Sunkel, O. and Gligo, N. (1980) *Estilos de Desarrollo y Medio Ambiente en América Latina*, (2 vols), Mexico: Fondo de Cultura Económica.

Sudjic, D. (2003) 'Talking Shop', *The Observer Review*, 31 August, p. 10.

Swyngedouw, E. (2000) 'Authoritarian governance, power and the politics of rescaling', *Environment and Planning D: Society and Space*, 18: 63–76.

Swyngedouw E., Moulaert, F. and Rodriguez, A. (2002) 'Neoliberal urbanization in Europe: large-scale urban development projects and the new urban policy', in N. Brenner and N. Theodore (eds) *Spaces of Neoliberalism*, Oxford: Blackwell.

Sykes, R., Bochel, C. and Ellison, N. (eds) (2001) *Social Policy Review 13*, Bristol: Policy Press.

Szul, R. and Mync, A. (1997) 'The path towards the European integration: the case of Poland', *European Spatial Research and Policy*, 4(1): 5–36.

Szymanska, D. and Matczak, A. (2002) 'Urbanisation in Poland: tendencies and transformations', *European Urban and Regional Studies*, 9(1): 39–46.

Tett, A. and Wolfe, J. (1991) 'Discourse analysis and city plans', *Journal of Planning Education and Research*, 10(3): 195–200.

Tewdwr-Jones, M. (2001) 'Complexity and interdependence in a kaleidoscopic spatial planning landscape for Europe', in L. Albrechts, J. Alden and A. da Rosa Pires (eds) *The Changing Institutional Landscape of Planning*, Aldershot: Ashgate.

Tewdwr-Jones, M. and Harris, N. (1998) 'The commodification of development control', in P. Allmendinger and H. Thomas (eds) *Urban Planning and the British New Right*, London: Routledge.

The Ecologist (1972) Blueprint for Survival, London: The Ecologist.

The Observer (2004) Climate change will destroy us Pentagon tells Bush, 22 February. 2004.

Thomas, M. J. (1996) *Planning and Radical Democracy*, unpublished paper, Oxford: School of Planning, Oxford Brookes University.

—— (1998) 'Thinking about planning in the transitional countries of Central and Eastern Europe', *International Planning Studies*, 3 (1).

—— (1999) 'The politics of environmental policy', paper presented at the *Futures Planning: Planning's Future* conference, 29–31 March, University of Sheffield, Sheffield.

Thomas, R. (2002) *Sustainable Urban Design*, London: Spon.

Thompson, J. B. (1990) *Ideology and Modern Culture*, Cambridge: Polity.

Thornley, A. (1991) *Planning under Thatcherism: the challenge of the market*, London: Routledge.

Tickle, A. and Welsh, I. (1998) 'Environmental politics, civil society and post-communism', in A. Tickle and I. Welsh (eds) *Environment and Society in Eastern Europe*, Harlow: Longman.

Todd, J. (1986) *Sensibility: an introduction*, London: Methuen.

Townroe P. (1996) 'Urban sustainability and social cohesion', in C. Pugh (ed.) *Sustainability, the Environment and Urbanisation*, London: Earthscan.

Underwood, J. (1991) 'What is really material? Rising above interest group Politics' in H. Thomas and P. Healey (eds) *Dilemmas of Planning Practice*, Aldershot: Avebury.

UN-Habitat (2004) *Commission on Sustainable Development, 12th Session* (opening statement and message from Anna Kajumulo Tibaijuka, Executive Director), www.unchs.org/csd/edstatement.asp (consulted 8 May 2004).

United Nations Conference on the Human Environment (1972), *Stockholm Declaration*, New York: United Nations.

United Nations Development Programme (2003) *El conflicto: callejón sin salida* (Human Development Report Colombia 2003), Bogota: UNDP.

United Nations General Assembly (1995) 'Revisión de las contribuciones a la implementación de la Agenda 21 sobre la acción nacional e internacional en el área de los asentamientos humanos', *Report of the General Secretary of the Habitat II Conference to the General Assembly*, A/CONF.165/PC.2/8, 13th Feb. 1995, Nairobi.

United Nations Commission on Development and the Environment for Latin America and the Caribbean, UNCDELAC (1990) *Our Own Agenda*, Mexico: IDB/UNDP.

United Nations Conference on Human Settlements (1996) 2nd, Istanbul, Turkey, *How cities will look in the 21st century*, Proceedings of Habitat II, Dialogue, Nagoya, Japan.

Urry, J. (1990) *The Tourist Gaze*, London: Sage.

Vattimo, G. (1988) *The End of Modernity*, Cambridge: Polity.

Vigar, G., Healey, P., Hull, A. and Davoudi, S. (2000) *Planning, Governance and Spatial Strategy in Britain: an institutionalist analysis*, London: Macmillan.

Viviescas, F. (1993) 'La calidad de la vivienda y la ciudad', *Revista Camacol*, 57: 73–89.

Walker, S. (2000) 'The West Midlands', in R. Simmonds and G. Hack (eds) *Global City Regions: their emerging forms*, London: Spon.

Ward, S. (2002) *Planning the Twentieth Century*, Chichester: Wiley.

Weltrowska, J. (2002) 'Economic change and social polarisation in Poland', *European Urban and Regional Studies*, 9(1): 47–52.

West Midlands Metropolitan County Council (1983) *West Midlands County Structure Plan: proposals for alterations, explanatory memorandum*, Birmingham: West Midlands County Council.

West Midlands Chief Officers Joint Working Group (2000) *Review and Assessment of Air Quality in Birmingham*, www.birmingham.gov.uk/environment (consulted 12 March 2004).

West Midlands Joint Committee (2003) *Moving With The Times: West Midlands Local Transport Plan*, www.westmidlands/tp.gov.uk (consulted 20 March 2004).

White, R. (2002) *Building the Ecological City*, New York: CRC/Woodhouse Publishing.

Wild, A (2000) 'Planning for sustainability: which interpretation are we delivering?', *Planning Theory and Practice*, 1(2): 273–84.

Williams, K. (2002) 'Measuring urban compactness in UK towns and cities', *Environment and Planning B: Planning and Design*, 29: 219–50.

Williams, K., Burton, E. and Jenks, M. (eds) (2000) *Achieving Sustainable Urban Form*, London: Spon.

Williams, R. (1977) *Marxism and Literature*, Oxford: Oxford University Press.

Wodz, J. (2002) 'New social partners of local power in Poland', *International Social Science Journal*, 172: 239–46.

Wodz, J. and Wodz, K. (1998) 'Environmental sociology in Poland and the ecological consciousness of the Polish people', in A. Tickle and I. Welsh (eds) *Environment and Society in Eastern Europe*, Harlow: Longman.

Wolaniuk, A. (1997) 'Spatial and functional changes in the city centre of Lodz', in S. Liszewski and C. Young (eds) *A Comparative Study of Lodz and Manchester: geographies of European cities in transition*, Lodz: University of Lodz.

Wolfensohn, J. D. (2004) *2004 Spring Meetings – Press Conference*, http://web.worldbank.org/WBSITE/EXTERNAL/NEWS (consulted 2 May 2004).

World Bank (1990) *Urban Policy for the 1990s*, Washington: World Bank.

World Commission on Environment and Development, WCED (1987) *Our Common Future*, Oxford: Oxford University Press.

You, N. (1996) 'Towards a habitable future' (interview), *City*, 3–4: 83–110.

Young, C. and Kaczamarek, S. (1999) 'Changing the perception of the post-socialist city: place promotion and imagery in Lodz, Poland', *The Geographical Journal*, Vol. 165, No. 2, 183–91.

Zetter, R. (2002) 'Market enablement or sustainable development: the conflicting paradigms of urbanization', in R. Zetter and R. White (eds) (2002) *Planning in Cities: sustainability and growth in the developing world*, London: ITDG.

Zetter, R. and White, R. (eds) (2002) *Planning in Cities: sustainability and growth in the developing world*, London: ITDG.

Index

Page references that include illustrations are in **bold**.

ESSENTIAL READING

Place, Identity, Participation and Planning

Cliff Hague and Paul Jenkins   Hb: 0-415-26241-0
Heriot-Watt University, UK   Routledge   Pb: 0-415-26242-9

Introduction to Environmental Impact Assessment

John Glasson, Riki Therivel
   and Andrew Chadwick   Hb: 0-415-338-36-0
Oxford Brookes University, UK   Routledge   Pb: 0-415-33837-9

Sustainable Urban Development: Volume 1

The Framework and Protocols for Environmental Assessment
Stephen Curwell, Mark Deakin
   and Martin Symes   Hb: 0-415-32214-6
University of Salford, Napier University
   and University of the West of England Routledge   Pb: 0-415-32219-7

Healthy Cities in Europe

Agis Tsouros and Jill Farrngton (eds)   Hb: 0-415-34001-2
The World Health Organisation   Routledge   Pb: 0-415-34002-0

Information and ordering details

For price availability and ordering visit our website www.tandf.co.uk
Subject Web Address **www.routledge.com**
Alternatively our books are available from all good bookshops

eBooks

eBooks – at www.eBookstore.tandf.co.uk

A library at your fingertips!

eBooks are electronic versions of printed books. You can store them on your PC/laptop or browse them online.

They have advantages for anyone needing rapid access to a wide variety of published, copyright information.

eBooks can help your research by enabling you to bookmark chapters, annotate text and use instant searches to find specific words or phrases. Several eBook files would fit on even a small laptop or PDA.

NEW: Save money by eSubscribing: cheap, online access to any eBook for as long as you need it.

Annual subscription packages

We now offer special low-cost bulk subscriptions to packages of eBooks in certain subject areas. These are available to libraries or to individuals.

For more information please contact webmaster.ebooks@tandf.co.uk

We're continually developing the eBook concept, so keep up to date by visiting the website.

www.eBookstore.tandf.co.uk

For Product Safety Concerns and Information please contact our EU
representative GPSR@taylorandfrancis.com
Taylor & Francis Verlag GmbH, Kaufingerstraße 24, 80331 München, Germany

www.ingramcontent.com/pod-product-compliance
Ingram Content Group UK Ltd.
Pitfield, Milton Keynes, MK11 3LW, UK
UKHW021830240425
457818UK00006B/147